Security Manager's Guide to Disasters

Managing Through Emergencies, Violence, and Other Workplace Threats

Security Manager's Guide to Disasters

Managing Through Emergencies, Violence, and Other Workplace Threats

Anthony D. Manley

CRC Press is an imprint of the
Taylor & Francis Group, an informa business

CRC Press
Taylor & Francis Group
6000 Broken Sound Parkway NW, Suite 300
Boca Raton, FL 33487-2742

First issued in paperback 2017

ISBN 13: 978-1-138-11369-5 (pbk)
ISBN 13: 978-1-4398-0906-8 (hbk)

Library of Congress Cataloging-in-Publication Data

Manley, Anthony.
Security manager's guide to disasters : managing through emergencies, violence, and other workplace threats / Anthony Manley.
p. cm.
Includes bibliographical references and index.
ISBN 978-1-4398-0906-8 (hardcover : alk. paper)
1. Emergency management. 2. Crisis management. 3. Industries--Security measures--Management. I. Title.

HD49.M362 2009
658.4'77--dc22 2009016220

Visit the Taylor & Francis Web site at
http://www.taylorandfrancis.com

and the CRC Press Web site at
http://www.crcpress.com

This book is dedicated to my wife, Emily,
and to the security professionals whose vocation entails the extensive
and specialized functions of protecting the total assets of the
institution or the business enterprise in which they are employed,
and to those public safety officers who heroically enter into disastrous
and extremely harmful situations where others fear to tread.

Contents

SECTION 2

SECTION 3

SECTION 6

Author's Note

Nothing should be construed as to how one should act or react based upon what is read or contained herein. The author accepts no liability for any type of damages, real, inferred, or imagined, or for any professional injury, personal injury, or property damage that might result from the use or misuse of any of the information, techniques, or applications presented or implied in this book. This book is intended as an educational and training publication and should not be considered a substitute for professional guidance or advice in the numerous emergencies that may occur. Moreover, concerning the laws or regulations that may govern the various incidents covered herein, consider consultation with your own attorney, as only an attorney can give legal advice. New laws are enacted routinely, and court decisions affecting new and current laws are handed down daily. The author suggests that you use this book in conjunction with legal advice from your counselor and the accepted procedures, requirements, and current law in the state in which you are employed.

The commentary, citations, and case law described in this book have been summarized in most cases to illustrate a point under examination and are taken from previous holdings. For a full review of a particular case, the reader should refer to the published case citation.

Regarding state citations noted herein and elsewhere, precedent may be offered to a court when a holding has been handed down on a previous similar case from another state, and only when that state has no case for precedent *may* the court defer to a decision from that other state. Federal holdings submitted or handed down to a state court will be binding on all states.

Note also that, throughout this book, the author has used *italics*, ***bold italics***, or **bold letters** for emphasis in order to highlight certain affirmations, narrative, and passages of importance to the reader.

Finally, and generally, the designation of loss-prevention manager will be used in this book, although the positions of safety director or manager as well as security or safety manager may also be found from time to time in the narrative. Nevertheless, whether each of these designations is in a management or an administrative position, all should be considered synonymous in meaning, authority, and title. Also, the reader should understand that the indication throughout this book regarding the terms *business, business enterprise, company,* or *institution* is the same in regard to the intent of the narrative.

Preface

In the grand scale of human events, we can safely say that, at this time in our history, the earth endures a human population that is more numerous, healthier, and affluent than ever before. However, because of overpopulation in many areas of our planet, we face an unprecedented risk of death and destruction caused by natural hazards such as earthquakes, floods, drought, famine, and wildfires. In addition, we also face major threats that are human-made, which include industrial explosions, major transportation accidents, mass violence, and our recent ever-growing exposure to terrorist activity. Moreover, because of our accelerating use (and misuse) of our natural resources and assets, our land, forests, and water are now considered to be at a greater risk than ever before. Notwithstanding that and along with the industrialization of the populace within the last 150+ years and its effect on the planet, we also have to contend with global climatic changes.

The term *disaster* can unquestionably be attributed to obvious geographical areas because of an exceptionally high concentration of people, poor economy, limited resources, and an inadequate or incompetent response of aid to the affected areas, whatever the cause. In addition, criminal behavior, accidental failures of design, or human mismanagement relating to large-scale structures, transportation systems, and industrial processes can cause serious loss of life and injury as well as property and environmental damage on a large scale. And never forget extensive epidemic and pestilence episodes, which in the history of mankind have caused death and illness of such magnitude that, in many cases, the scale of the event could not be measured.

At this point in our historical record, we have to ask how we can contain or prevent large-scale disasters. There is no doubt that a major effort must be initiated by nations united against the most recent threats, particularly terrorism and environmental and climatic changes. Moreover, principal corporations acting globally in concert must become involved more deliberately in assisting individual governments for this common purpose.

So the question is, with prior knowledge or not, how can we prevent, control, or at least alleviate a serious occurrence that could affect our lives, health, business, and environment?

In order to answer this question—while recognizing that including every type of disaster or catastrophe in one book would entail an enormous task—we can nevertheless address the more prevalent disasters that the midlevel and administrative loss-prevention and safety executives would normally encounter, and thereby expose them to situations that they can prepare for in a timely and judicious manner. In that regard, this book is offered to those security and safety professionals employed in the realm of risk management and loss prevention for their consideration and for the possible enhancement or modification of their emergency procedures.

Introduction

Realize that no disaster or emergency can be accurately forecast, and no amount of preparation can anticipate every detail that will or might occur. It would be impossible to anticipate and prepare conclusive disaster control instructions on a nationwide basis, since every facility is distinct to some degree in its mission and operation. Moreover, depending on the incident that a particular facility may face compared with what another may encounter elsewhere in the country, plans will probably differentiate as to details. This could include the type and effectiveness of the civil emergency response, the severity of the incident, the time of day, the size of the workforce and occupants on board at the time of the incident, and the emergency equipment and supplies on hand and immediately available. Additionally, security officers will be dealing with situations that could bog down the emergency response, such as traffic control and anxious and agitated people in a highly stressed state of mind, causing obstacles that must be dealt with over and above the original incident. Therefore, loss-prevention and safety personnel must be prepared to adjust or deviate from any written plan upon direction from appropriate authority. However, we must consider that in the absence of this authority, the security officer should have the ability to make immediate rational decisions during the occurrence that will be based on his or her training, experience, and expertise.

A disaster or serious emergency can occur in many forms; it can be sudden and without warning, such as a terrorist act or an earthquake, or it can have some forewarning, such as severe environmental factors or observable employee behavior prior to a serious assault upon others. Generally, though, other than natural occurrences, we become complacent and many times indifferent to indicators that our business may suffer damage or destruction, or that employees and the public we serve may become victim to death and injury.

Our society relies on its businesses to generate jobs, provide a tax-based revenue, and to nurture a built environment that is healthy and sustainable. When a business protects itself from natural, accidental, or in some way, human-made disasters, it also protects one of its community's most valuable assets. Bear in mind, therefore, that the best time to respond to a disaster is before it happens.[1]

Consider that our country, like many other countries, is vulnerable to international and domestic terrorism. Anyone, anywhere could be the subject of a violent terrorist act, whether the perpetrator is foreign or native to our shores. Along with that, any business establishment or institution may be subject to acts of violence by a disgruntled or deranged individual, where the inhabitants may be killed or seriously injured, and/or where extensive property damage could result.

Moreover, to a great degree, incidents of this type cannot be anticipated. But whether the act is planned or random in nature, business executives and loss-prevention managers can employ measures to prepare for, respond to, and help prevent death, injury and damage in some measure. The American public must forget the mindset that it can't happen here or it won't happen again. Recent attacks on the World Trade Center in New York City, the Pentagon in Washington, D.C., the Oklahoma City bombing of the Murrah Federal Building, and the Columbine High School and Virginia Tech tragedies show that it can happen and most probably will happen again.

"Since 2001, the business world has been rocked by an unprecedented series of catastrophes, such as the September 11 attacks, accounting scandals, and Hurricane Katrina to name just three."[2] If nothing else, these events have elevated the concerns of those corporate managers and administrators in regard to risk management on how to lessen known and unknown exposures, internal and external threats against property, assets and people, and the protection of corporate integrity.[3]

We must also consider the distinctions between catastrophes that are natural, accidental, or human-made, and the ever-growing emergence of technological and engineering development. One is the distinction between the promotion of technology or scientific study and its control. Other considerations are natural or human-made disasters, which technology might prevent, or those catastrophic risks brought about or made more dangerous by technology.

Modern science and technology have enormous potential for harm, but they are also abundant sources of social benefits.[4]

The Loss-Prevention Professional

Today, more than ever, the assets of a company are at risk, not only from terrorist activity and natural or accidental occurrences, but also from criminal and civil liability in its various forms. Anything that will reduce the value of the company by loss of revenue, loss of monetary capital and holdings, and loss of customer and investor confidence must be considered the responsibility of that person hired to protect all of the assets of the business enterprise. This individual may be classified by various titles, but all have the fundamental duty of *asset protection*. These titles may include, risk manager, loss-prevention manager, security director or manager, safety director, or fire safety director. Depending on the extent of the business establishment, he or she must have the experience and expertise, along with the necessary resources, to effectively and efficiently accomplish his or her responsibility. For our purposes, we will refer to this able and competent individual as the *loss-prevention manager.*

The professional who holds the position of loss prevention, safety, or risk management has access to many references, texts, and manuals regarding safety and security. Unfortunately, there is no one concise manual that encompasses all of the more serious emergencies, threats, or disasters that could affect a business enterprise into one easily read text for the business practitioner whose vocation is to protect the assets of the company. Moreover, as far as the author is aware, there is no formal training that completely encompasses the varied occurrences or threats that a private security and safety manager may have to face and/or his or her immediate response to those incidents. Academia is, however, slowly catching up by engaging and expanding security administration and homeland security programs, among others of that genre. In the public sector, particularly relevant concerning the emergency services rendered, training is an ongoing process with scenarios and strategies conducted for proficiency and safety of not only the public they serve, but also the public safety personnel. Training under near-actual conditions is expensive and, therefore, it is only in rare instances with very large corporations in the private sector that loss-prevention personnel will have access to training equal to that of professional public safety services.

Law enforcement and other public safety agencies have the responsibility, resources, and ability to effectively mobilize and respond to a disaster or emergency, no matter what type or how serious. The private sector is much less trained and organized to respond to any type of serious emergency or disaster. Generally, loss-protection personnel in the private sector have little or no training or preparation in the area of serious emergencies, other than myriad short courses or seminars that they may be able to attribute to their profession. Unless one is trained and exposed to the various emergencies and crises that would encompass or involve the training by a municipal or government entity, his or her entry into the field of private loss prevention lacks that exposure.

Along with the objective of emphasizing the role of the loss-prevention professional, this book hopes to underscore another viewpoint: the aftereffects of an incident in the form of liability. Frequently, in that sense, personnel in the public and private area have little knowledge of *both* criminal and civil liability. To be inclusive in this narrative, it is necessary that the author must cover, to some degree, the types of liability usually encountered.

Liability Issues

The purpose of explaining in some depth the various criminal or civil liabilities that an individual and/or a business may face is an effort to instill in the reader the fact that today we live in a litigious society where excessive monetary awards for damages, real or imagined, have become all too common. Business owners and management are increasingly held liable for not making their premises safe for employee, customers, visitors, students, and/or patients. Potential areas of security and safety issues and of workplace-violence-related litigation that should concern employers include civil actions for negligent hiring, worker's compensation claims, third-party claims for damages, invasion of privacy and civil rights actions, and Occupational Safety and Health Administration (OSHA) violations, charges, and fines.

If the awards granted to the plaintiffs are high, and many times exceptional, the insurance rates will be affected by that rise. Ultimately, as with any loss, the cost of doing business will also rise, and in effect the higher cost of products and services will be borne by the public. Many times, the costs over and above liability coverage will affect the bottom line of the business along with the goodwill and stature that a business has built up over time. The knowledge and involvement that a loss-prevention manager has concerning all issues of corporate liability, security, and safety will be considered an important asset to his or her employer.

We must be cognizant of the fact that, during any normal business operation, emergencies do occur and the possibility of damage, injury, or death could affect the company's various assets, including the goodwill of that company if response to that emergency is slow, apathetic, or negligent. A terrorist or violent act is not the only course of action that can cause a disaster of enormous proportions to a business endeavor. The loss-prevention professional must be aware that the safety and security of people and property are not the only considerations; there is also the whole consequence of monetary loss due to damage, destruction, injury, death, and liability issues to the company and its occupants. Moreover, this will include the business that is partially or completely shut down and could, without a doubt, affect the bottom line to the point that it becomes a disastrous situation for the business enterprise.

Preparation and Confrontation

Prevention may be the most difficult objective related to a terrorist or violent act. In reality, we are dealing with an unknown and, many times, an unpredictable escapade. Because of the nature of the business establishment, in some instances where access by the public is the norm, it becomes much more difficult to prevent unlawful acts in the workplace. So then, how do we attempt to minimize criminal acts against our facility and prepare for those occurrences that may be unavoidable?

The answer may be as simple as *knowledge*, *preparedness*, and *training*.

The objective of this book is to identify those factors or conditions that could result in a disaster or any emergency in or near an institution or a business establishment. In addition, it is designed to familiarize those loss-prevention professionals employed to protect those business assets about the what, when, and how to react in the event of such an occurrence.

In essence, this book is designed to help the reader to effectively identify, analyze, plan, protect, and respond to an emergency occurrence, and to recognize the ramifications of the aftereffects upon a business enterprise.

This manual is not written to replace those public safety measures set in place by the professionals responsible for the safety and welfare of the public and community at large. Moreover, this manual is not intended to present detailed procedures or programs already offered or mandated by federal, state, or local emergency management authorities. It is written for the loss-prevention professional whose duty is to safeguard the assets of the company where he or she is employed. These assets include the personal safety of employees, customers, visitors, patients, and/or students; the personal and corporate property of the company or the institution. He or she must prevent the possible disruption of the business enterprise and its operation; and last but not least, maintain its goodwill as perceived by the public.

If nothing else, what is presented here will have served its purpose in its attempt to expose the varied disasters and emergencies that could occur and to provide a comprehensible framework for identifying, analyzing, preparing for, and responding to these occurrences.

Additionally, this book is intended to apprise the loss-prevention manager of the fundamental policies and procedures of disaster and emergency management at an actual and immediate event. It is an attempt to acquaint the loss-prevention manager with what may occur in various incidents, what is expected of him or her in safeguarding life and property, and what must be done to achieve the recovery or the restoration of the business enterprise.

Nevertheless, as we have noted, disasters come in many forms. Whether the cause is a terrorist or criminal attack, fire emergency, civil disruption, storms and floods, geographical disruptions, or a major electrical failure, the loss-prevention professional must be aware of how such a crisis would affect all that he or she is employed to protect and know how to prepare and confront each emergency.

Most importantly, adequate procedures must be planned prior to any event that may cause a danger or possible harm to business assets. This will include appropriate safeguards, resources, and training of personnel to effectively and efficiently rise to the occasion.

This book will attempt to encompass the more important emergencies that may confront the private loss-prevention manager and staff in the protection of people, property, the prevention of crime, and the avoidance of civil and criminal liability.

In that regard, we will discuss the following objectives:

1. Prevent or reduce the severity of the incident.
2. Initiate immediate and professional response to reduce loss of life, injuries, property damage, and loss.
3. Provide for adequate interaction and cooperation with all public safety agencies, local government, and other public and private utility services.
4. Provide a system for recovery and restoration by limiting the losses and returning the business to operation as soon as practicable.

SECTION 1

Chapter 1

Loss Prevention

The Objective for Security and Safety

Natural, accidental, and human-made disasters have inflicted monumental and severe costs in human and physical resources, including damage to the environment, the American psyche, and our way of life. Considering the effects of these offensive acts on the public good, the incidents and conditions described in this manual represent potentially significant obstacles not only to the economic growth and development of a geographical area, but to the business community in particular. In this regard, we hope to cover the more significant disasters, emergencies, violence, and threats within a business enterprise.

Disaster Defined

The generally accepted definition of a disaster is an event that disrupts the daily life of a population within a community or a large geographic area and that could result in substantial loss of life, serious injury to many of the populace, and social upheaval. Disasters or emergency situations may include or be caused by natural, accidental, or human-made occurrences. Such occurrences can leave many injured, homeless, jobless, helpless, and hungry. The disruption, destruction, or loss of vital public services such as water, electric power, communications, and transportation could have a long-lasting effect on a community's infrastructure during a disaster of catastrophic proportions. When the impact on a community is so great that the available resources cannot cope with the occurrence effectively, outside help is needed. But we must realize that federal, state, and local public safety agencies may be overwhelmed or have been affected to the extent that only emergency services can be rendered during the initial stages of the emergency.

With that in mind, any threat, emergency, or serious occurrence within the business enterprise may be considered a disaster when it has a critical effect on the operation of the business or institution by causing a significant business disruption or a complete shutdown. Therefore, we may consider that a disaster in the business setting is an unplanned occurrence that disrupts the routine operation of that company. Business disasters come in many forms and, other than natural occurrences, may also include fire, terrorism, workplace violence, robbery, burglary, computer crashes, sabotage, espionage, or any type of an incident that causes a serious business intrusion or disruption. We must also consider that, along with physical damage to life and property, a business enterprise may also be seriously affected monetarily by the awarding of damages via civil litigation by persons who have been harmed in some way. Businesses have been known to fail completely because of high awards over and above any liability coverage they might have had.

According to an annual security forecast survey conducted soon after 9/11 by *Security Magazine* for the year 2002, the top ten concerns facing businesses at that time were:[5]

1. Terrorism
2. Employee theft
3. Computer security
4. Property crime
5. Access/egress
6. Violent crime
7. Vandalism
8. Burglary
9. Parking lots/garages
10. Liability insurance/white-collar crime (tie)

In this study, two-thirds of the respondents said that recent terrorist acts spurred them to reevaluate their security programs. In fact, terrorism was not even listed among the top ten security concerns for the year 2001.[6] In the years 1999 and 2000, security executives rated workplace violence as number one in the top ten security threats and issues facing corporate America.[7]

Loss Prevention Defined

Today, *loss prevention* is considered a broad term regarding the protection of a company's assets. Prior to our present period, the terminology of security and safety were separate disciplines in which security encompassed the protection of one's assets from theft or intentional damage, the identification of the perpetrator, and the apprehension of that person. Safety, in general, included the protection of people and property from unintentional personal harm, damage, or liability. The term *loss prevention* combines both disciplines into one application and function.

The present loss-prevention manager, safety, or security officer employed in the preservation of company assets must not only be aware of duties and responsibilities for the protection of persons and property from harm, but also of the possibility of criminal or civil litigation caused by acts or omissions committed by an outside individual or group, by oneself, an employee, or employer.

The Role of Loss Prevention

The primary role of the loss prevention department is the protection of life and property. To effectively accomplish this, Loss Prevention must examine, identify, prepare, plan, test and manage the components of security and safety.[8]

This will include protection of all persons in, on, and, oftentimes, around the business property, in which the security and safety are delegated to one department or group. In effect, anyone on the premises will fall under that umbrella of protection. Moreover, the assets of the company, such as merchandise, stock, supplies, proprietary information, machinery and equipment, the facility itself, and the company's good will, require considerable effort of protection from loss in any manner.

Additionally, other than the protection of life and property, we must consider the effectiveness of reacting to an emergency, recovery procedures, and backup capabilities for the resumption of the business.

The Role of the Loss-Prevention Manager

The loss-prevention manager wears many hats. Depending on the type of business and the facility to be protected, the loss-prevention manager's duties and responsibilities may encompass the following:

Risk identification

- Identify threats, risks, or possible losses that may affect the business

  - Review and evaluate present policies, procedures, resources, and other pertinent sources and data
  - Review current safety and security procedures in an effort to ensure an effective response
  - Determine that the business has adequate and applicable insurance coverage (adequate in the sense that all risks are covered, no matter what the event; depending on the geographical location, this would include earthquake, flood, hurricane, and tornado insurance)
  - Identify the potential economic losses from various occurrences
  - Evaluate the available equipment and the outside support services as well as the required and necessary actions of key personnel during an emergency incident
- Development and implementation
  - Develop policies, procedures, and specific programs to eliminate or effectively respond to causative factors or incidents
  - Train the personnel who may have to respond to a specific threat or incident
  - Establish the required actions, operations, notifications, and responsibilities when reacting to a specific threat or incident
  - Establish the procedures for cleanup, recovery, and return to full business operations
- Testing and measurement
  - Develop testing and measurement procedures to determine the effectiveness of a policy or program
  - Assign priority to all programs involving safety and security training, testing, and inspection over any other instruction or workshop
  - Test and drill the appropriate personnel on a routine basis, and document the results of such activities

The Necessary Attributes of a Loss-Prevention Manager

The loss-prevention manager must have the necessary qualities to effectively perform in this vital and demanding role, particularly in a large business establishment. The following subsections outline the favorable attributes that a corporate owner should expect in the selection of a loss-prevention manager.

Expertise

The loss-prevention director/manager should have

- A working knowledge of local, state, and federal laws as they pertain to the business of security and safety

- A good working relationship with local law enforcement, public service and safety entities (including the local fire department, fire inspectors, and building code enforcement officers), and prosecuting attorneys
- Direction and close control over the compilation and maintenance of pertinent security and safety records (Reports should be accurate, complete, and neatly retained for reference or presentation, as required for criminal or civil litigation or as obligated by law.)
- The ability to professionally interact with insurance investigators and attorneys representing both the plaintiff and defendant and, in particular, the attorneys who represent the loss-prevention manager and the company
- Experience in court appearances and at various pretrial hearings, especially with regard to personal deportment, evidence control, and presentation
- The ability to train and instruct subordinates and other employees on relevant issues concerning safety and security
- Extensive loss-prevention experience or, at the least, appropriate certifications with three years of prior experience in security and/or safety as a manager or supervisor (The candidate should not come directly from a law enforcement position unless he or she has an exceptional background and knowledge of the required tasks and functions in security and safety as a director or manager.)

Attitude

The loss-prevention manager should

- Not have a "cops and robbers" mentality
- Emphasize prevention, as opposed to detection, as the most effective means of loss prevention
- Comprehend that the powers of investigation, arrest, and apprehension are very limited compared to that of a sworn police officer

As to the three aforementioned approaches, the author has found that police or law enforcement officers entering the private sectors of loss prevention, security, or safety generally have a mindset conditioned by their many years of exposure to crime and punishment. Essentially, this will encompass a *response* to an incident, determining whether an offense has occurred, and if so, arresting the perpetrator—a reactive rather than a proactive perspective.

Completing the above list, the loss-prevention manager should also

- Be even-handed, fair, and honest
- Recognize that, without the cooperation of all employees, any loss-control program would be unproductive (In that regard, an ongoing training pro-

gram for all employees and an effective safety committee must be considered a necessity.)

- Have open and frequent communication with all administrative, management, and subordinate employees

Management Ability

The loss-prevention manager must

- Be able to balance the problems and priorities of the business in relation to the safety and security of all assets, and to investigate, find, and direct practical solutions that affect the loss-prevention department's mission and those of other business areas and personnel
- Have the ability to effectively present or communicate ideas or proposals to upper management and the company administration (The thoroughness and thoughtfulness devoted to all relevant business issues in presenting these thoughts should gain the respect and approval of management.)
- Be a good manager and trainer, but most of all, a motivator of people
- Have the highest standards of honesty, integrity, and veracity, and should expect the same from subordinates within the department

Responsibility

The loss-prevention manager should not consider his or her job or department as a separate entity from all other managers and departments within the business establishment. The loss-prevention manager should recognize that the facility manager or CEO has the primary responsibility for loss prevention, and that the loss-prevention manager in reality is responsible to that superior in the following ways:

- Control of programs, contractual services, and procurement, including security hardware evaluation, procurement, installation, and operation
- In charge of a proprietary security guard force, or has complete autonomy and control over the procurement, selection, proficiency, placement, and discipline of a contractual guard operation (This will include the complete control and supervision of contractual security services such as extra guards for specific purposes, undercover operations, and alarm maintenance and service.)
- Examination, identification, analysis, and correction of security and safety problems that affect the assets of the company
- Development of guidelines, policy, and training (Consider that if a guideline is important enough to be written and is to be adhered to in the strictest sense, it is policy. *Policy is not policy unless it is written.*)

- Develop guidelines and training for security officers in investigation, apprehension, arrest, and prosecution of offenders (This will include instruction regarding trespass laws, security devices [locking apparatuses, closed-circuit TV, electronic article surveillance], basic first aid, cardiopulmonary resuscitation, and security officers' specific job tasks and functions.)
- Take responsiblity for conducting investigative procedures, correlating and determining fault, and setting up corrective action concerning safety and security issues
- Develop standards and guidelines to protect facility, stock, merchandise, equipment, and most importantly, the protection and preservation of proprietary information and property
- Achieve compliance with federal, state, and local laws and regulations regarding safety and security
- Develop guidelines for training of all managers and employees regarding their actions concerning various security, safety, and liability issues
- Develop policy regarding health, safety, and legal issues, both criminal and civil
- Act as the motivator for open communication between all managers and employees relative to security and safety issues and the protection of the company's assets

Evaluation

However the performance appraisal or evaluation procedure is promulgated by company management, the following criteria may be considered in the evaluation process of a loss-prevention manager along with other analysis that may pertain or is relative to that particular business:

- How the loss-prevention manager and his or her subordinates are supported and perceived by company administration and other management, supervisors, and employees
- The number of new policies and procedures presented by the loss-prevention manager that are accepted and implemented by upper management
- The reduction or the avoidance of potential damages, injury, or civil liability litigation brought against the company and its employees for acts committed, omitted, or perceived
- The investigation and case management of all civil and criminal litigation brought to bear against the company
- The reduction of employee injuries and lost worker-hours compared with prior years

In regard to crimes attempted or committed on or around the facility, the following statistics may be of concern in the evaluation process:

- The number of lawful apprehensions of customers or visitors committing unlawful acts on or within company property (The standards and percentages should be based on the performance of comparable similar business establishments.)
- The number of internal investigations as opposed to cases still active or cases closed by:

Arrest	Restitution
Recovery of property	Termination of employment
Civil recovery	Other

Conclusion

We can safely say that the loss-prevention manager has the responsibility of protecting the employees of the company in which he or she is employed and the public who may be on, in, or around the facility from any risk that can cause harm.

We can also say that this manager will, at times, be at odds with senior managers or administrators when enforcing safety and security rules or the expenditure of funds to install, upgrade, or reinforce programs, procedures, or equipment. Because the scope of the job description for a loss-prevention manager is very wide, this individual will become involved in all aspects and all areas of the company. And because this will lead to interactions with superiors, he or she may be viewed as an irritant or, in many instances, simply ignored. It is therefore incumbent that the loss-prevention manager have the authority to enforce the company's written policies and procedures regarding all safety and security issues. It must be determined how sincere the company is in its view of safety and security, and what the effect will be on a department head, supervisor, or the employee who refuses or is lackadaisical in responding to a company rule or regulation, whether flagrant or minor.

In many instances, the administration or management of a company may view the loss-prevention department as *reactive* rather than *proactive*, in that they may only observe or recognize actions by security officers responding to a specific incident. In an effort to circumvent this viewpoint, the loss-prevention manager must maintain and routinely provide to the administration reports and records of statistics, audits, inspections, investigations, and studies, among others, to nullify any adverse perception by management.

Chapter 2

The Emergency Procedure Plan

Unless a particular emergency or disaster must be accounted for as a specific incident, the management and resolution of any emergency can be simply established in an Emergency Procedure Plan.[9]

Every business or institution that employs several people and has, as part of its normal business, the coming and going of visitors, customers, clients, patients, or students should have an Emergency Procedure Plan that could be put into effect by knowledgeable people in any emergency. Naturally, the larger the facility and the more substantial the number of workers and inhabitants, the more complex the plan will be.

This plan may also be designated as the Emergency Safety Plan or the Emergency Preparedness Plan. Such a plan will fix responsibility, indicate those employees who should be accountable, and detail the procedures to be taken during the various emergencies or business disruptions. The introduction to this plan will essentially make clear a ***policy statement*** regarding the company's or the institution's commitment to safety and security.

The company may compile separate strategies that are specific in nature to the occurrence. These plans may include a disaster plan (or Disaster Procedure Plan) and a Fire-Safety Plan. If the business establishment is considerable, both in size and occupants, it is suggested that separate written plans regarding disasters and fire be considered to specifically categorize assignment, response, and responsibility. In any event, any plan of this nature must be attached to and made part of the Emergency Procedure Plan.

Every loss-prevention department must consider emergency response as one of the most important functions that will be expected of its staff.

Emergency Planning

The plan must be in writing. It must be precise and specific in nature, defining in terms of organization the normal-to-emergency modes for each incident that may occur. The plan is developed to prioritize the following goals: protect and save lives, minimize injuries, protect property, reduce liability, and diminish the exposure of a shutdown. Following these considerations, the plan should provide restoration procedures.

Once a crisis or an emergency situation occurs, the two key elements of the plan are

What to do, and who should do it!

Planning for Emergencies

The following assessments should be taken into consideration:

- Determine the company's objectives
- Consider the hazards; define in terms that are relevant to the organization
- Determine the present or existing safeguards
- Determine the need for new or more effective safeguards
- Prepare written procedures (Emergency Procedure Plan):
  - Specify the occurrence
  - Specify those employees who will respond to a particular incident
  - Specify proper notifications to public safety agencies, public utilities, support services and company management
  - Institute safety training for all personnel and special training for specific safety and security personnel
  - Specify the drills or testing of procedures on a routine basis
  - Revise, review, and revise

Content of the Written Plan

The plan should describe in detail the following areas of concern:

- The company's policy
- Risk assessment

- Description of each emergency or serious incident that could occur along with details regarding the procedures for each, particularly during and after the emergency (Depending on the facility, subdivision plans such as fire-safety, disaster, or bomb-threat procedures could be separate, attached to, and made part of the Emergency Procedure Plan.)
- A list of all emergency equipment available on site
- The assignment of the person (or alternate) who will be in charge of the emergency (or the incident) and the location of the command center
- A list of all employees who are first-aid responders, members of the fire brigade, and all personnel who are to respond depending on the particular incident
- A list of all emergency mutual-aid and contractual services (police, fire, ambulance, telephone, electric and gas, heat/cooling systems, elevator repair, alarm repair and maintenance, glass replacement, property recovery, recovery and restoration, etc.)
- A listing of emergency telephone numbers of all management and administrative personnel of concern who are to be notified depending on the incident
- Evacuation and shutdown procedures
- Identification of the proper authority that will determine when reentry to the facility may occur after an evacuation
- A definitive plan or procedure of moving from normal operations into and out of the emergency mode of operation

The plan should detail every emergency that could occur within the facility, which would include but not be limited to fire, gas leaks, bomb threats, explosions, natural disasters, blackouts and brownouts, and even lost or found children (if applicable to the premises).

The plan should also include the following important considerations:

1. In what manner will the incident be relayed to the facility's occupants and, in particular, the specific emergency responders?
2. What are the location and responsibilities of the fire command station (which may be part of and/or located at the disaster command center)?
3. What decisions and notifications are to be made? When? By whom?
4. What are the responsibilities and duties of the person who will be in charge of the command center?
5. What duties or responsibilities are assigned, and to whom?

6. Who should respond to the emergency or serious incident? When, where, and how?
7. Who becomes "in charge" at the scene of the incident and will be in direct contact with the command manager?
8. Who will make public announcements or act as the spokesperson if the media become involved?

Because ***company policy is not policy unless it is written***, a thorough and detailed written emergency procedure plan will serve as a written directive and guide for all employees in an emergency. In particular, it will be a guide for the loss-prevention/security officer, who will be part of the response team in most, if not all, emergencies.

The loss-prevention manager should review, revise, or upgrade the emergency procedure plan on a regular basis. All security personnel should review and examine this plan also on a routine basis so as to reinforce the required actions or responses expected of them, and to become acquainted with any new procedures.

Following this segment, the reader will find a listing of emergencies and disasters that could occur at, near, or within a business establishment. However, in this book, we hope to cover in some detail the more significant disasters, emergencies, violence, and threats that may occur and seriously affect an enterprise.

The Emergency Response Team

The employees noted below may be considered part or all of the emergency response team, depending on the facility and the type and severity of the emergency they are assigned as responders.

In any business environment the Emergency Procedure Plan will describe duties assigned to knowledgeable personnel who will be a part of the emergency response team:

- The loss-prevention manager, the safety or the security manager, or the risk manager
- The fire-safety director or safety director

- The maintenance manager or the building engineer of the facility
- All security officers and security guards
- First-aid responders (may include security officers)
- All employees assigned to the fire brigade (may include security officers)
- All maintenance employees or engineers on duty
- The building administrator, facility manager, or store manager
- The operations manager or assistant store manager
- The duty manager or the senior manager for the day
- Members of the computer response team
- Any other employee whose expertise or position is applicable to the occurrence (EMTs, volunteer firefighters, etc.)

The loss-prevention manager must consider that emergency management and emergency response are required parts of the job description. He or she therefore must be the most important individual on site during a disaster or emergency. If he or she is unable to be present during such an incident, the immediate supervisor or deputy to the loss-prevention manager must assume control.

Contingent on the organizational makeup of the company, some of the above managers or employees may also be first-aid responders or part of the fire brigade. Also, some of the above persons may not be at the work site or on duty at the time of a fire or other serious emergency, but an adequate number should be present so that the necessary action may be taken. Whether the security officer is proprietary (employed by the company he is protecting) or contractual (employed by a guard company contracted and assigned to the company he serves), the officer must be considered part of the emergency plan, wherein his or her duties, tasks, and functions are specified.

The security officer must remember that, in any emergency, the primary concern is the safety of life and limb, followed by safety of property. The specific response by the security officer will depend on the type of emergency and what duties and responsibilities he or she is authorized and required to perform. The key point to remember is that the security officer must act in a professional manner. Likewise, the security officer who has been trained and is knowledgeable as to his or her duties in responding to different emergency situations is an asset to the employer. The officer's actions and demeanor will have a calming effect on all those he or she will come in contact with: supervisors, coworkers, and building inhabitants. If the officer panics or becomes excited or agitated in reaction to or during the course of the emergency, both employees and others will view the occurrence as more serious than it may be and, if nothing else, will become overly concerned, fearful, and possibly panic.

The security/loss-prevention manager and all security officers within the loss-prevention department should be aware of their actions and responsibilities during incidents such as those described on the following pages. It is therefore incumbent for the company or institution to have a written Emergency Procedure

Plan in effect that details what is an emergency, how it is to be handled, and who does what and when. First-aid responders and fire brigade members should have training and certification within their scope of duties. Additionally, all employees of a business establishment, no matter how large or small, should have training in fire safety soon after being hired, with subsequent adequate yearly retraining. This should also include the required hazard communication and bloodborne pathogen training mandated by OSHA if they are so noted to be included under those standards.

Protective and Emergency Equipment

Depending on the personnel assigned to a particular response, but in particular all security and safety officers, everyone should have certain protective gear available in emergencies. As an example this could include

First-aid kits, complete with the necessary equipment for the treatment of general injuries
Hard hats, eye protection, gloves
Self-contained breathing apparatus
Hand-held radios
Appropriate flashlights
Pocket knives, whistles
Larger emergency-type tools (described below)

Be mindful that comprehensive first-aid kits should be easily accessible and located at various locations throughout the facility. Larger implements such as Haligan tools (for forcible entry), pick axes, and Beil tools (personalized multipurpose forcible entry and rescue tools) should be centrally located for easy access in an emergency. An excellent location would be the disaster/fire command station. In addition, all security and safety responders should possess master hard keys and card keys for access to all offices, public areas, utility closets, equipment rooms, and warehousing and storage areas.

Cautionary Note

Depending on the laws of the municipality or state in which the security officer is employed, the use of noxious agents (chemical spray devices, including mace, pepper spray, tear gas), batons (nightsticks), or restraints akin to handcuffs and plastic ties may be unlawful or restricted in some manner. Concealable weapons such as electronic stun guns, blackjacks, slapjacks, and billy clubs are definitely unlawful to possess. Check with local law to determine what constraints are particularly directed toward security officers.

In Summary

Does your business have an Emergency Procedure Plan, and does it cover all types of emergency incidents? Does your business or department conduct and support the following ongoing procedures?

1. Be self-critical on all proactive and reactive strategies.
2. Is the plan up-to-date?
3. Is the plan reviewed routinely for necessary changes, additions, or deletions?
4. Fix what does not work.
5. Change what is necessary.
6. Train, drill, and inspect.

In summary

SECTION 2

Chapter 3

Specific Threats and Emergencies

Threat Assessment

In assessing safety and security, we must recognize that no conceivable likelihood of harm or havoc can be overlooked. There is, however, the concept of *relative risk*. As the principal of a loss-prevention department, your responsibility is *risk management*—the ability to foresee a risk and then to prepare or protect against that risk, or at least to reduce the losses that will accompany the event.

Because the terrorist threat is constant and most probable in some way or form at this time in our country's history, and in view of the risk of the enormous numbers of deaths and injuries that can be caused by a single incident, we will begin our narrative and cover this topic in some detail.

Consider the following in assessing a terrorist threat against the American populace, its government, and its institutions. Surprise and the method of attack are the greatest factors in causing uncertainty, fear, panic, and a reduction in free movement and achievement. Large cities, landmarks, historical and governmental sites, and attractive venues that draw large crowds will attract a terrorist bent on creating a serious attack.

All civil authorities, federal and state governmental agencies, and private security and safety alliances and enterprises must be constantly aware that, at this time in our history, we must be proficient and equal to the task of a perpetual threat assessment against those who would harm us. In that manner, note the following major areas of assessment:

1. Because of our large population and diverse ethnicity, the terrorist can easily blend in.
2. Because of our large population, which attends, works, and travels in a variety of areas, including prominent public events, major tourist sites, and public transportation in its various forms, the terror effect along with mass panic will be greatly enhanced.
3. The potential for an attack within a large group will greatly increase deaths and injuries.
4. The attack will cause the municipality's personnel resources—police, fire, emergency medical responders, health-care workers, etc.—to become overwhelmed as time goes by.
5. Depending on the type of attack, mass exodus from the area by people and vehicles may cause roadways and bridges to become jammed as people panic in an effort to escape from the initial engagement, causing emergency responders to be restricted in their efforts of assistance and response.
6. Depending on the weapon or type of attack, the infrastructure may be greatly affected. Damage to water, natural gas, and electric utilities; bridges and tunnels; and major thoroughfares requiring reconstruction and/or decontamination will place a burden on the municipality.
7. Fear and angst will be greatly intensified as the media, particularly television, cover the event, causing further panic, while also giving the terrorists what they want—publicizing the event and their cause.

A nuclear device would have a low priority because of the sophisticated manner, techniques, cost, and expertise required to produce one. Similarly, a biological or chemical attack would require a sophisticated manner of dispersal. Consequently, most experts agree that the most easily placed weapon is a bomb that can be transported or carried into the location of detonation. Whether the bomb is carried into an area by one or more suicide individuals or transported by auto or truck would depend on the target and the amount of damage intended. Consider a bomb that contains both explosive material and radioactive elements (radiological dispersal device) that, when detonated, releases not only the bomb blast, but also subsequent radiological contamination, which would be the predominant effect desired, as it would cover a larger area.

Chapter 4

Critical Business Threats That Cause Emergency Situations

The loss-prevention specialist must be prepared for any type of incident—probable or possible—that could cause death, injury, damage, or disrupt business operations.

Although we will cover the more serious threats to a business enterprise, any action or threat that has an effect on the bottom line of the business in question must be considered for an immediate reaction by risk-management or loss-prevention personnel. Minor events might include internal or external larceny, fraud, intentional or unintentional damage, equipment failure, and strikes or actions by employees that cause a loss.

The occurrences, threats, or disasters noted below may occur at any time and are not necessarily noted in the order of importance of the threat to the business community and public safety.

The loss-prevention manager as well as the security and safety officers must be knowledgeable of and follow the *emergency procedure plan*, which should describe the procedure that he or she is to follow for the various incidents that may occur at the company or institution to which he or she is employed or assigned.

The disastrous or emergency situations identified in the following chapters are considered to be serious enough to specify in some detail. We will begin with *terrorism* in Chapter 5.

Chapter 5

Terrorism

The Growth of Terrorism in the United States of America

On February 26, 1993, an explosion occurred in the World Trade Center, New York City. The explosion caused six deaths, 1042 injuries, and nearly $600 million in property damage.

On April 1, 1995, an explosion occurred at the Alfred P. Murrah Federal Building, Oklahoma City. The explosion caused 168 deaths, more than 400 injuries, and … "losses from the bombing, including lost business opportunities and unreimbursed medical expenses, would total $651 million."

In the five year period of 1989 to 1993, the Bureau of Alcohol, Tobacco and Firearms (ATF) has reported that nationwide there were 7,716 bombings, 1,705 attempted bombings, 2,242 incendiary bombings, 4,929 recovered explosives, and 2,011 hoax devices. Although the ATF information is considered highly representative and sufficient to permit valid chronological, geographical and trend analysis, it should not be considered all-inclusive; more incidents occurred than those reported through national data collection.[10]

On September 11, 2001, the greatest national tragedy that our country has ever faced occurred when hijacked commercial aircraft crashed into the Twin Towers of the World Trade Center in New York City, with total destruction of the site, and the Pentagon in Washington, DC, along with a total loss of lives in excess of 3,000.

Public Awareness

The odds of being a victim of a terrorist attack are remote unless you live in a large populated metropolitan area or those venues already noted. However, the dangers of a nuclear, biological, or chemical attack have to be addressed realistically, because at this time in our history, an attack from within or from without is a real possibility.

During this period, terrorist experts believe that an attack by an offensive nuclear device is remote, but not unlikely. It is believed that the most likely objective would be a vulnerable target attacked with conventional weapons such as car or truck bombs. The greatest risk is faced by the major metropolitan areas are regions with large populations.

Previous attacks on Oklahoma City, New York City, and Washington, DC, caused fear and anxiety among the American populace, at least initially. In the past few years, America has become complacent and lethargic. Security and safety enterprises grew exceptionally following 9/11, but as time passed, American business lost the desire to maintain these private sectors of safety. Constraints of money, time, effort, expertise, and an outright lack of commitment have put homeland security on the "back burner."

Areas such as maritime shipping and terminals, major landmarks and stadiums, public facilities, rail, and in general public transportation still lack or generally fail in the most rudimentary measures of security. Terrorist experts believe that another serious attack against our country is forthcoming and are waiting for "the next shoe to drop." Until average Americans believe they could be victims of a terrorist act—no matter where they live or work—only fear and concern will drive the desire to energize the fight against terrorism.

Terrorism Defined

Terrorism may be defined as any overt act by a person or a group that causes death, bodily injury, and destruction. Moreover, the act instills fear, panic and apprehension to the public at large, which ultimately disrupts the normal working environment of a business, a government agency, or the community in general.

The United States Department of Defense (DOD) defines terrorism as "the calculated use of violence or the threat of violence to inculcate fear; intended to coerce or intimidate governments or societies in the pursuit of goals that are generally political, religious or ideological."[11]

The bomb is the favorite device used by the terrorist. A bomb is easy to make, can be easily transported, can be concealed in various shapes and sizes, and can be easily placed within or near a building, structure, or a populated area. It is loud and it is very destructive, and therefore serves the purpose intended. By not recognizing innocent people who may be killed and injured, the terrorist has an infinite number of targets of opportunity, and this range of targets gives the terrorist an exceptionally high probability of success with little or no risk.

A terrorist attack against a business or any location may take several forms. It will depend on the technological means available (nuclear, chemical, biological, etc.), the nature of the issue (whether political, religious, or otherwise), and the various points of weakness in the terrorist's target.

As in any criminal act, the performance of the terrorist will depend on the desire, the means, and the opportunity.

Types of Terrorist Incidents

The Prevalent Threat

The terrorist act is rooted in promoting a particular lifestyle, religion, political viewpoint, or a bias against some other group. The aim is not only to cause death and destruction, but also to draw attention to the terrorist's ideology and philosophy. The threat of terrorism is becoming more prevalent. According to federal and national police agencies, terrorist attacks are likely to come in the following forms:

Nuclear, radiative, chemical, or biological attacks: Directed toward extensive areas or large groups of people, causing death, injury, and damage and resulting in a complete stoppage or slowdown of the social order and structure. This must be considered the most serious threat to public safety.

Bombings (in various forms): Conducted to bring about as much destruction and injury as possible. Bombing could include a nuclear threat, as noted above.

Hijackings and kidnappings: Usually directed toward high-ranking politicians and business executives. It may include assassination or a particular demand after the abduction.

Murder and assassination: Usually directed toward a targeted politician or business executive at any location convenient for the act.

Examples of Possible Terrorist Incidents

In an article dated March 17, 2005, *Newsday* covered an internal report, put out by the Department of Homeland Security (DHS), that outlined several hypothetical attacks. Once classified as secret, the report came to light after it was accidentally posted on the Web by the state of Hawaii and reported by the *New York Times* on March 16. Chagrined by this early discharge, DHS immediately released the report to help local, state, and federal officials prepare for the real circumstance. Some of the scenarios are described in Table 5.1.

Table 5.1 Hypothetical Attacks as Outlined by DHS: Weapons, Scenarios, and Estimated Casualties

Weapon	*Scenario*	*Casualties*
Nuclear device	Uranium stolen from former Soviet republics is used in bombs and detonated in a U.S. city	Can vary widely
Anthrax	Anthrax sprayed from a vehicle driven through large cities causes infections, many of which are fatal	13,000+
Plague	Toxin released in bathrooms of airports, sports arenas, train stations, enclosed spaces or conveyances	10,000+
Blister gas	Toxin sprayed from an aircraft over a sports stadium or other large event, causing burns and possible deaths	70,000 injuries
Sarin	Toxin released into ventilation/heating ducts in office buildings or enclosed public conveyances, killing up to 95% of occupants	6,000+
Toxic chemicals	Petroleum facility is attacked with rockets and/or bombs; fire and blasts release toxic plumes	1,000+
Chlorine tanks	Attack at an industrial facility causes release of gas, producing skin burns, inhalation injuries, and contamination of water	17,000 deaths

As in any large disaster, immediate deaths and long lasting injuries will only add to the initial event as fear and panic increase and public mobility comes to a standstill.

Targets of the Terrorist

In addition to the type of attack, we must be cognizant of a threat or act toward a particular target, which could include public transportation and buildings, for the purpose of promoting the terrorist cause. The following list identifies potential terrorist targets (not necessarily in order of concern):

1. Tourist regions, parks, stadiums, or other areas where large crowds will be found
2. Major governmental or historic monuments or landmarks
3. Buildings housing federal or state agencies
4. Large buildings with a large capacity of tenants and visitors, i.e., buildings that are of national or international significance
5. Major utilities: electric, telecommunications, gas, and water
6. Nuclear and chemical plants, gasoline refineries, and oil and gas storage farms
7. Aircraft, airlines, rail lines and equipment, bridges, subways (underground and underwater), and other significant transportation facilities
8. Maritime transportation and cargo operations at major ports; includes container transport as the dominant concern
9. International business establishments or businesses that have large holdings in another country
10. Universities, colleges, and research laboratories
11. Abortion clinics or similar service providers
12. Courthouses, town halls, schools, churches, police stations, post offices, and tourist and foreign trade offices
13. Political leaders and principal business executives, including their families, personal residences, and work sites

The Classification of Weapons of Mass Destruction

The following description and arrangement of weapons of mass destruction has been created by the author to clarify the severity of the two and to make the distinction between the two categories more easily understood.

First class—Atomic (nuclear), radiological, biological, and chemical weapons or devices that can be utilized in various ways and that, when activated, can cause serious and profound damage to our country's financial, economic, commercial, and industrial infrastructure, in addition to illness, injury, and death to thousands upon thousands of the populace.

Second class—Weapons that can have a severe impact, depending on their destructive force and location at the time of activation. This class would include mechanical and chemical bombs of all types and treacherous letters, packages, and other mailings with intent to cause death or injury.

Chapter 6

Weapons of Mass Destruction of the First Class

Threat of an ABC Attack

The occurrences identified here include atomic (nuclear and radiological) attack as well as biological and chemical attack (ABC attack). They also include attacks upon the commerce of a country, institution, agency, or business, whether conglomerate, national, or international in nature. Moreover, the aggression can occur in many ways, and it need not be terrorist driven. Further discussions on these subjects are noted later in this book.

Other than the terrorist's use of a car or truck bomb, or the nefarious use of a smaller bomb device against a particular group or location, we must envision other acts that threaten our way of life. In essence, the following must be of great concern because of their effect on life and death. They may be considered as weapons of mass destruction, not only within the business facility, but some distance from that facility, and within reach of the community at large. Accordingly, serious consideration must be given to where a known or perceived threat might affect the business operation.

In that regard, there are three types of weapons that can cause cataclysmic effects on the populace in a given area. These can be described as atomic, biological, or chemical (ABC) threats and are distinguished as follows:

1. ***Atomic (nuclear) devices***: The most extreme devices for death and destruction over a large geographical area, with the subsequent exposure from radiation to a large population; capable of causing severe and long-lasting injury
2. ***Biological methods***: The introduction of biological pathogens (bacteria, viruses) into the air, water, or food supply or via other manufactured products that are consumed or otherwise come into contact with the public
3. ***Chemical agents***: The use of poison gases or other toxic agents that can cause harm by inhalation, ingestion, or topical deposition on the skin, particularly in closed areas such as buildings, theaters, subways, aircraft, etc.

The characteristics of the above weapons are described in the following sections.

Atomic (Nuclear and Radiological) Attack

Nuclear Attack

Nuclear weapons represent the most extreme destructive devices and the ultimate weapons for causing severe and extensive damage to people, businesses, institutions, and property over a very large area. They inflict severe damage over a large area that will cause massive injury and death along with radiological contamination producing acute and chronic radiation sickness. The results can include complete devastation, mass slaughter and injury to humanity, radiation sickness, and fallout onto other areas. The nuclear weapon must achieve a chain reaction with material that is very hard to acquire or accumulate. The making of a nuclear device requires a sizable infrastructure and scientific know-how. Moreover, it requires acquisition of fissionable material that is under close scrutiny by the international community.

- If a terrorist uses such a weapon, the effects would be catastrophic and devastating. The following conditions would take place:
  1. The initial blast would cause a severe shock wave, static overpressure, and wind that would flatten everything and kill all humanity in an extensive area.
  2. Intense heat and light (thermal radiation) would cause fires that would blind and burn people.
  3. Radioactive fallout would cause radiation poisoning or long-term sickness over a large area.
  4. The blast could cause an electromagnetic pulse (EMP) that could destroy any unshielded electronic components in vehicles, computers, radios,

and telephones. (See the section about electromagnetic pulse devices in Chapter 7.)

- Experts agree that a nuclear device would not be the weapon of choice for several reasons:
  1. The cost and technology involved to construct a nuclear bomb would be beyond the abilities of a terrorist organization.
  2. The requirement of plutonium and uranium needed for the bomb is, again, beyond the capabilities of the terrorist.
  3. Unless a rogue national entity has the time, money, and scientific capability to produce an atomic bomb for its own purposes, the present consensus is that the terrorist will find and use a more easily acquired weapon.

Radiological Attack

Radiation Poisoning

Since 9/11, the media has alarmed the public with scenarios of nuclear radiation that would have catastrophic consequences. In reality, radiation poisoning depends on the total exposure dose. Radiation will vary according to the yield of the weapon, the distance from ground zero, the amount of protective shielding, the length of exposure, the amount of radiation received, and individual health. Radiation sickness is not contagious, and there is no treatment other than supportive care.

Power Plant Attack

Because there is a history of the use of aircraft in terrorist attacks, it is not inconceivable that a plane or planes could be used to fly into a chemical plant (see discussion in the chemical attack section later in this chapter), an electrical plant, or a nuclear power plant. Nuclear plants present a particularly enticing target to aircraft or bomb-laden trucks breaching fences or barriers to cause the dome or the reactor to collapse and emit radioactivity into the air surrounding the plant and ultimately into the atmosphere. In this scenario, radiation poisoning would take place over a large geographical area.

Biological Attack

Biological Contamination

Biological contamination involves the introduction of living agents (bacteria, viruses) into the air, water, or food supply or into manufactured products that are consumed by or otherwise come in contact with the public. Biological agents can be defined as infectious microbes, organisms, or toxins that can produce illness or death in people, animals, or plants and that can be dispersed as aerosols or airborne particles. Because these agents are not easily detected and there is a time element before detection, the terrorist could contaminate food or water, with the public completely unaware of the attack until it is too late. Biological agents include anthrax, tularemia, plague, smallpox, and the Ebola virus in its various forms. Some agents may be contagious. Contamination always requires immediate medical attention.

Some biological agents are more difficult to manufacture into weapons than other agents, or to disperse them into the air or food supply to any great degree, but they have the potential to inflict widespread damage.

Small-scale attacks have occurred in the United States in the recent past. In 1984, at The Dalles, Oregon, a religious cult laced salad bars in several restaurants with salmonella. Approximately 750 people were exposed and became sick. In 2001, exposure to anthrax contained in mailed envelopes caused five deaths.

Water Supply and Air Contamination

The widespread dissemination of a strong biological agent into our water supply or into HVAC (heating, ventilation, and air conditioning) systems throughout the business world would be a cause for alarm. Casualty numbers would be immense because a biological attack makes use of a covert weapon that acts slowly, several hours or days after the initial dissemination. Infection would be unknown until people began to show symptoms, and it would be expected that metropolitan areas would be targeted and suffer the greatest casualties.

The availability of antibiotics and vaccine will depend on the stockpile of these supplies and the speed of delivery to first responders in the target area. The distribution of medicine and the initiation of decontamination procedures will likely overwhelm the civil authorities as they attempt to service the populace.

Bioagents cannot be self-treated, as contamination requires a medical regime. Effects can include severe sickness and death. Dissemination of these agents can be made by aerosol sprays, gas, and liquid, any of which could be delivered by various mechanical means, including bombs and rockets.

Agroterrorism

Another area of concern is a biological attack causing deliberate contamination of our food supply, particularly at its source—the farmland of our country. Moreover, contamination could be created during shipment to the marketplace, in distribution centers, during preparation or packaging, or while in storage. We have a complex system of food distribution, and a terrorist could strike anywhere within that system. A biological agent would be the most likely choice of dispersal. If a chemical agent is used, the concept would be to destroy crops, thereby causing a shortage in the food supply.

Pandemics

Pandemics have caused millions to become ill or die. The viruses that have caused the World War I variant of the flu or our present-day AIDS (HIV) virus are further discussed under natural and common environmental occurrences in Chapter 11. However, a pandemic need not be caused by natural occurrences preying on the human condition; it can also be initiated intentionally by groups intent on furthering their cause or ideology. Consider the possibility of a pandemic caused by fanatical members of various organizations serving as carriers of a deadly disease such as smallpox, who enter into crowded areas and infect others, who in turn will infect many hundreds more. Once the capability to produce an agent is achieved, large quantities can be manufactured and stored for future use. Anthrax or smallpox would be sufficient for the terrorist to use, rather than a hybrid-type weapon with no defense, which, if released, would put the whole world at risk, including the terrorists and their cause.

Biological Agents

The three basic groups of biological agents that would likely be used as weapons are bacteria, viruses, and toxins.

1. ***Bacteria***: Small free-living organisms that reproduce by simple division. If a bacterial disease is introduced into the populace, it often responds to treatment with antibiotics.
2. ***Viruses***: Organisms that require living cells in which to reproduce on the body they infect. Viruses produce diseases that generally do not respond to antibiotics, although drugs are sometimes effective.
3. ***Toxins***: Poisonous substances found in and extracted from living plants and animals. Some toxins can be produced or altered by chemical means. A few toxins can be treated with specifically selected drugs.

Bioagent Categories

Devised by the CDC, bioagents are categorized in three risk groups:

1. *Category A*: Includes agents that have the potential to create the greatest number of casualties; most likely to be deployed by terrorists and may include anthrax, smallpox, bubonic plague, botulism, tularemia, and hemorrhagic fever
2. *Category B*: Includes agents that, although serious, would not cause stress to the civil authorities and the employment of a health-care system as would occur if a category A dispersal took place
3. *Category C*: Includes emerging agents

Not all biological agents are contagious. For example *Variola major*, the causative agent of smallpox, is extremely communicable, while *Bacillus anthracis*, the causative agent of anthrax, is negligible in its ability to be transmitted between individuals.

Chemical Attack

The use of poison gases or other toxic agents can cause harm by inhalation, ingestion, or topical deposition on the skin. An attack of this type can be particularly effective in closed areas such as buildings, theaters, subways, aircraft, etc. Some agents are intended to incapacitate only, while others tend to cause immediate death. Early indications must identify the type of weapon utilized, how much of a warning we will receive, and what areas are targeted or contaminated.

Chemical agents may consist of poisonous gases, liquids, or solids that can kill or incapacitate people or destroy livestock and crops. Chemical agents can be odorless and tasteless and, as in the case for biological agents, are difficult to detect. The severity of the injury will depend on the type and amount of the agent as well as the duration of exposure.

The use of chemical agents differs from a bombing, and a chemical attack can involve one or more chemical compositions or mixtures, as noted in the following narrative.

Public Transportation and Enclosed Structures

Chemical formulas and specifications are, in reality, commonly known within the chemical and biological professions. A competent terrorist trained in how to handle and disperse the agent can do severe harm to the populace. The agent may be in a liquid or gaseous form and may be deployed in many ways. The ideal location for such an attack would be a closed or semisealed area such as a subway or a public building. The heating, ventilation, and air-conditioning (HVAC) systems can spread the agent throughout the building in minutes. Results in an open area would not be as effective, since weather or wind conditions would disperse the chemical in an uncontrolled manner.

Chemical Plants

Chemical plants and depositories are ideal sites for accidental discharge, incompetent control, or a terrorist attack. Compared with nuclear plants, these areas are poorly protected and, at times, carelessly controlled. Ammonia and chlorine, two commonly produced commodities, can cause serious injury and death if leaked into the air in their gaseous form, where wind can disperse them over a wide area. Most chemical plants are located near urban areas and are likely targets for the terrorist.[14]

In April 2007, the Department of Homeland Security (DHS) initiated a new Chemical Facility Antiterrorism Standard (CFATS) that was subsequently published in the *Federal Register*. This law requires chemical facilities fitting a certain profile to complete an on-line risk assessment, and those companies determined to be at greater risk will be required to conduct vulnerability assessments and submit site security plans that meet DHS performance standards.[15] Although many chemical companies voluntarily instituted security programs, DHS concluded that these voluntary efforts alone would not provide sufficient security for the nation. This standard was signed into law in October 2006 by the President of the United States as the Department of Homeland Security Appropriations Act of 2007, which gave DHS the authority to regulate the security of high-risk chemical facilities. In December of that year, prior to its emplacement, public comment was sought, and many state and local governments publicly denounced several sections of the standard as too weak. They also believed that, under this law, the federal government would have the ability to overrule state laws on chemical security and, in fact, weaken those state laws already in operation. Notwithstanding some states' dissension over this law, CFATS compliance regulations were put into effect. Special requirements included training of all security officers in the chemical and petrochemical industries in safety awareness, OSHA safety topics, hazard recognition and response, and accident investigation and analysis. In essence, the law mandated education and training geared to the chemical industry.[16]

Chemical Agents

In an effort to acquaint the loss-prevention professional, a basic understanding of those chemical agents that have been and can be used against humanity are described here. Noxious chemical agents include the following:

Blister agents: Includes mustard agents, lewisite, and phosgene oxime. Usually deployed as a gas and principally designed to incapacitate, affecting the skin, eyes, and the respiratory tract. There is no known antidote.

Mustard agents: Although mustard can be deployed as a liquid, gas seems to be the choice, affecting the eyes, skin, and respiratory tract. It causes red and itchy skin, painful eye irritation, vomiting, abdominal pain, bleeding nose, and coughing of blood. Can be fatal if exposure is concentrated. There is no known antidote.

Lewisite: Contact will cause an immediate skin irritation and blisters. Symptoms are similar to exposure to mustard agents. However, lewisite can also affect the cardiovascular system, causing a shock to the body and creating a low blood pressure. Antidote (British anti-lewisite—BAL) should be administered as soon as possible after an exposure.

Phosgene oxime (CX): An agent that almost immediately causes pain. The agent can permeate clothing much faster than any other blister agent. Causes skin blanching, hives, and formation of scabs, compounded by intense itching and severe pain. There is no known antidote.

Nerve agents: Includes tabun (GA), sarin (GB), soman (GD), and VX. These are the most probable agents of choice for the terrorist, as they are fast acting and lethal, with VX the most deadly. A lethal dose of nerve gas can kill in 15 minutes; in liquid form, death will occur within 2 minutes. When inhaled as a vapor, and depending on the concentration, the victim will suffer eye, nose, and lung irritation; tightness in the chest; and the gasping for air. Finally, the victim will pass out and suffer paralysis, with all body systems ceasing, and death will occur. Upon skin contact, these agents may take a few minutes more before the symptoms noted above will appear. (If exposure occurs, immediate action would be to self-administer into the victims' fleshy part of the thigh an auto-injector of atropine.) Nerve agents of this type, when utilized in a confined space, can be very effective. Closed areas such as subways, tunnels, theaters, or public conveyances containing large groups of people would allow the agent to produce the desired maximum effect.

Germany accidentally discovered nerve agents in the 1930s when scientists were researching new and improved insecticides. The technology was subsequently provided to the military; tabun was the first agent developed, followed by sarin.

Blood agents: Include hydrogen cyanide (AC), heavier than air, and cyanogen chloride (CK), lighter than air. A lethal dose of a blood agent will cause cellular asphyxiation, resulting in increased respiration, seizures, cessation of breathing, and finally cardiac arrest.

Toxins: A poison created from living organisms such as bacteria, plants, fish, algae, fungi, and other forms. Usually deployed in the form of an aerosol,

> these agents are technically categorized as a chemical weapon, since they poison the victim rather than infect. Botulinum toxin is considered the most poisonous toxin known to humankind. At this time, it appears that municipal chlorination or application of other disinfectants in drinking water are incapable of destroying this toxin.

It is expected that if a chemical attack does in fact occur, the civil authorities will set up mass decontamination stations in the affected area, along with medical services as needed. Of course, depending on the type and strength of the agent or agents released into the target area, decontamination results will be questionable. For further information about all acceptable protective measures and antidotes that a safety or loss-prevention professional may procure, see Appendix A.

Conclusion

Chemical and biological agents can be dispersed in the air we breathe, the water we drink, the food we eat, and through contact with domestic animals. Dispersion methods can be as simple as opening a container or using an aerosol as a scattering instrument, or it can be as elaborate as detonating an improvised explosive device.

Chemical agents, however, are more easily produced and dispersed and, therefore, are probably the terrorist's first choice. Based on present technologies, the employment of chemical weapons can cause the death of tens of thousands of people, and the use of biological weapons could kill hundreds of thousands of people. Readily available commercial materials and technologies found at pharmaceutical and chemical plants or fly-by-night laboratories can produce these agents with ease.[16]

For a detailed examination of the three types of ABC weapons of mass destruction described in this chapter, see Appendix A.

Chapter 7

Weapons of Mass Destruction of the Second Class

Bombs and Bomb Types

Other than the more serious weapons of mass destruction (WMD) of the first class discussed in Chapter 6, the bomb—in whatever form—must be considered one of the more dangerous implements of destruction. The basic types of bombs are identified in the following subsections.

Mechanical Bombs

Mechanical bombs use triggering or detonating devices to cause an intense buildup of heat, gas, and inside a container until it explodes or shatters. The container may be filled with a chemical explosive such as dynamite, TNT, nitroglycerine, mercury fulminate, black powder, or a mix of common fertilizer and fuel oil.

Small Bombs

A small bomb can be very powerful and can easily be concealed in a backpack, handbag, briefcase, or shopping bag. It may be dropped off at a particular site or carried into a target area strapped to the body of a terrorist willing to lose his or her life in order to complete the task. Although bombs of this type may be considered small, they contain assorted metal fragments such as nails, bolts, etc., so as to cause injuries that are more serious. A terrorist setting off a bomb in a mall or a market place containing a multitude of people can cause a severe amount of death and injury.

A small device may also be disguised as a pack of cigarettes or a hand-carried radio, and although it may seem to be diminutive, the consequences of a detonation can be grim. In addition, letter bombs or bombs contained in small packages are not uncommon and can cause severe personal injury or death to recipients.

Car Bombs

A large bomb can be concealed in any type of vehicle and, when detonated, can cause complete devastation to anything in the vicinity. The ingredients or makeup of the bomb have no significance as long as the device will be explosive enough to complete the task. Any vehicle operated by a person determined to achieve this type of destruction would be difficult to deter unless known beforehand. The most popular technique to reduce damage and injury is the installation of barriers around a building so that a vehicle would not have the opportunity to park or enter too close to the structure. Vehicles that enter underground or interior parking or delivery areas must be subject to close observation and strict control of access and egress. Of course, the type of vehicle control will depend on the business endeavor and the extent of the possible threat.

Incendiary Devices

These devices are normally designed to ignite after a business has closed for the night, so that once the fire has started, discovery is not immediate. The usual scenario is where the perpetrator slips or places the device into the sides of upholstered chairs and sofas. Retail stores that carry furniture of this type are at great risk, but it should be remembered that such a device could be placed anywhere, so that all businesses are susceptible. Loss prevention must consider that it is almost impossible to prevent delivery of a device of this type onto the business premises. If an incident such as this happens once, it can be expected to happen again. At times where a high risk is anticipated, security officers may be required to conduct as complete a search as soon as possible after the business has closed for the evening.

Radiological Dispersal Devices

This type of device is commonly known as a ***dirty bomb***, which in effect is a conventional explosive packed with radiological waste. Although a dirty bomb makes use of radiological material, it is not a nuclear device. A bomb of this type will use its radioactive component in the form of a solid, liquid, or powder, and when the explosion occurs, it will disperse a low-level radiological contamination. Contamination by radiation is activated by a mechanical device that sets off a high explosive (basically a mechanical bomb, as described above) in order to disperse the radioactive material. A device of this type will not be as destructive as an atomic bomb, since it will be only be as destructive as the size or amount of the explosive. The aim is for the conventional explosive to disperse the radiological material contained in the bomb container.

The initial blast will most probably kill dozens, but the radiation will affect many more. Emergency responders arriving at the scene will be unaware that they will be exposed to radiological poisoning. Moreover, nuclear waste, unlike fallout from a nuclear explosion, does not decay rapidly, and the radiation levels take years to deteriorate rather than days. Depending on the wind and other environmental factors, the dispersal of radioactive material would cause the land to be contaminated and unusable for many years, and adversely affect a large portion of the populace with immediate and long-term radiation sickness.

Once this type of weapon is known and has been deployed by a terrorist, the general public—not knowing when the next incident will occur—will live in fear and anxiety, since this type of weapon is easy to produce and transport.

Electromagnetic Pulse Devices

We cannot rule out another explosive device that is not widely known outside of military and law enforcement. An article in the Technology Section of *Newsday* on March 9, 2005, "A Potent Threat—Feeling More Vulnerable," written by Deborah Barfield Berry, attached to the Washington bureau, gave the world a wake-up call regarding an electromagnetic device not known to the general public.

Fundamentally, it is a bomb that can cause the complete failure and shutdown of every electrical or electronic component or device within its range of operation. Such a device is considered by many as a technological incursion into our present-day reality from science fiction.

There is a possibility that terrorists may have the capability to manufacture or have access to a primitive version of a high-tech weapon that previously was limited only to governments, and if terrorists do not yet have access to this technology, they certainly will in the near future. The purpose of this weapon is to unleash an ***electromagnetic pulse attack*** that would cripple the nation, damaging electrical systems vital to everything from financial services to transportation and communications. Every system and infrastructure dependent on electricity and electronics could be sufficiently harmed to qualify as catastrophic to the nation.

Experts have stated that the United States is particularly vulnerable because it relies heavily on telecommunications and electronics, which are integrated into practically every business and personal system in the country. Consider that an attack of this type would disable financial and defense systems, electrical power stations, all vehicles, marine and air transport ignition systems, and all radio, television, satellite, and telecommunications. Such an attack could result in a complete stoppage of business, movement, and activity of any kind. Consider the effect on the populace, as panic and fear grow with the increasing disruption of all services.

A basic or primitive version of the weapon can be simply described as a bomb made up with batteries as the generating device, and an explosive encased in a copper tube, which is completely surrounded by coiled copper wire. This bomb can be built from common electrical parts and explosives. The basic firing sequence is described as follows:

1. Electric current in the coil creates a magnetic field.
2. The explosive is triggered at one end.
3. The explosion races down the tube, compressing the magnetic field as the coil bursts, causing a powerful wave of magnetic energy to penetrate walls, concrete, and most metal shielding.

However, a more sophisticated device based on the initial elements described here would be a much more effective weapon over a larger area.

While not a new threat, since a pulse attack was a major concern during the Cold War, the 9/11 attack on our country has heightened this threat to a more serious concern. A security panel warned Congress of the seriousness of this threat to communications, transportation, and banking on March 8, 2005, a day before the aforementioned *Newsday* article appeared. Lowell Wood, acting chairman of the Commission to Assess the Threat to the United States from Electromagnetic Pulse Attack, a section of Homeland Security, and Peter Vincent Pry, a member of the commission, stated clearly that the nation's telecommunications systems are vulnerable. Both noted that potential adversaries and terrorists obtain nuclear weapons to generate the electromagnetic impulse. At one point in the hearings, the 2003 blackout on the East Coast was noted as an example of the devastating impact of losing power. Much of this report has not been made public because of security concerns.

The panel also advised Congress that other countries are beginning to address this threat, which is considered an indication that this issue has broad international credibility.

Chemical Bombs

The explosion of a chemical bomb is caused by an immediate and rapid conversion of a solid or liquid substance into a gas that has much greater volume. Bomb contents such as fertilizers and other ordinary products are readily available, easily purchased, easily transported in bulk, and, at present, not regulated in any way.

In the Oklahoma City tragedy at the Murrah Federal Office Building, the bomb that was contained in the truck used by the perpetrator was determined by the FBI to be made of ammonium nitrate–based fertilizer combined with fuel oil and primed with a detonator that, when set off, caused a powerful explosion. The result was almost complete devastation to the building and its occupants. This combination of ingredients was also used in the first attack on the World Trade Center in New York City and in the bombings of two U.S. embassies in Kenya and Tanzania.

We must consider that a terrorist may have possession of and use any one of these implements to further a cause. Whatever the design, any one of these devices will cause death, great harm, destruction, a destroyed or weakened infrastructure, and create fear, panic, and apprehension in the population at large.

Suspicious Packages and Mailings

This is an area of ongoing concern that is addressed on a daily basis. Every day, somewhere in the United States, suspicious packages or letters are received by the U.S. Postal Service, U.S. Customs, government agencies, or nongovernment institutions and business enterprises. Many times they are found to be bogus, even though they may resemble a mail bomb or contain materials that appear to be biological or chemical in nature.

Parcel and Letter Bombs

These devices may be contained in large or small boxes or oversized envelopes. In order to avoid over-the-counter post office inquiry or to prevent a package or envelope from being returned because of insufficient postage, perpetrators commonly affix excessive postage for the size of the package. These bomb devices are set to go off when opened, with the intention of injuring or killing the recipient. Letter bombs may appear to be rigid, uneven, or lopsided. Package bombs may have soft spots and bulges or be irregular in shape. In either case, the appearance of oil or grease stains observed on the wrappings should be a possible indication of a bomb within. The most recent extraordinary case of bombings was that of the "Unabomber," later identified as Theodore Kaczynski, who killed 3 people and injured 23 others all over the United States from the late 1970s to the early 1990s. Although he was found to be mentally ill, other perpetrators include terrorists and those with criminal intent.[17]

Biochemical Letters

The use of letters containing a biological agent led to heightened awareness when several letters containing anthrax spores were received during two separate periods. The first occurred over several weeks beginning in mid-September 2001, shortly after the 9/11 attacks. The first occurrence included letters sent to several media organizations, TV and radio news stations, and newspapers. Five letters were believed to be mailed at that time, and the first person to die from exposure to anthrax worked at a Florida tabloid called *The Sun*. The second set of letters containing anthrax was mailed on October 9 and directed to two U.S. senators. More lethal than the first group of letters, they contained nearly pure spores of anthrax. In these latest incidents, at least 23 people developed anthrax infections, with 18 considered to be life threatening. Five people ultimately died of inhaling the anthrax powder. The anthrax spores in the first attack appeared to be a coarse brown granular material. The spores in the second attack were more potent—a highly refined dry powder consisting of nearly pure spores of anthrax. Anthrax is a powdery substance, typically brownish in color, and resembles a very fine cocoa.[18] (See Anthrax in Appendix A.)

Because of these incidents regarding explosive and biological agents, and the chance of injury or death and infection by touch or inhalation, government offices and large corporations have set precautionary measures and guidelines in their mailrooms.

For further details and procedures in handling articles of this type, see the subsection entitled "Threats or Bombs Received by Mail or Messenger" in Chapter 7.

Chapter 8

Combating Terrorism

Blocking the terrorist act is difficult at best unless one has confidential intelligence or inside information that could effectively deter the event. Other than that, resisting the terrorist in any threat or encounter involves two sets of actions:[19]

Antiterrorism involves the use of *defensive measures* to (a) reduce the vulnerability of people and property and (b) enhance response to and containment of the event. Terrorists rely on surprise, minimal security measures, and confusion at the time of the incident. Effective security equipment, procedures, and personnel could essentially deter a terrorist act, and along with proper reactions and response can reduce the opportunity to act or at least reduce the drastic results that could occur.

Counterterrorism involves the use of *offensive measures* to neutralize terrorist groups, prevent, deter, and respond by national directives and actions. The overt and covert use of the military and national agencies to render the terrorist or terrorist group extinct constitutes counterterrorism.

At the present time, the protection of water treatment stations and reservoirs, tunnels, and transportation facilities and conveyances from an ABC (atomic, biological, or chemical) attack is almost nonexistent. Large businesses, particularly in fully occupied high-rise structures, may have little or no safeguards whatsoever. A bomb or a chemical or biological attack at any of these locations could kill and injure thousands.

Because an attack that includes the devices or schemes described above would be without warning, the only action to take would be to recognize the type of threat and react accordingly. Once the loss-prevention professional recognizes that a person or persons are becoming ill or reacting to an unknown agent, medical attention must be forthcoming. Regarding a biological or chemical exposure, medical and public safety authorities must determine the cause and the method of treatment, and at this point proper caution, procedure, and care will be disseminated. In an atomic blast, regular bomb procedures would take effect.

The above types of occurrences will most probably occur surreptitiously by an international or a domestic terrorist acting on behalf of a cause. We must also bear in mind the possible actions of a deranged person who, for whatever reason, may wish to cause serious harm or death to a large segment of the general public. In any event, the public emergency services—local, state, and national—will become involved and control the incident based on the extent of the threat or activity.

In particular, we must not forget the extremist carrying out radical themes, someone who is willing to give his or her life without hesitation for a cause. An attack of this sort is extremely hard to deter, since the terrorist gives little or no thought to personal safety. The use of a nuclear or incendiary device must also be considered, since the terrorist will make the choice and regard the performance and the outcome equivalent to meeting his or her goals, no matter how much death and devastation will result.

The placing of a bomb within a building or in a crowded locale so as to cause physical damage, personal injury or death, panic, and fear has become the norm in many Third World countries, particularly in Europe and the Middle East. Recently, our country has also become the target of national and international radicals who are intent on inflicting damage to life and property in both business and government organizations. Consider also the public apprehension and fear that follow such an incident. The disruption of a business is a natural development following a bomb explosion, causing people to be fearful of using public transportation or entering into public areas, where they may expose themselves to possible harm.

We have seen terrorist activity increase dramatically in recent years, and the United States has not been immune to such attacks, whether the terrorist was a national or international entity. Indeed, our country has become one of the most targeted by terrorists. Besides patronage and humanitarian assistance, the United States has financial and military interests globally that have placed many of its citizens in geographic areas that are politically unstable. Many countries despise our lifestyle and way of life, our concept of religious freedom, and our interference within their borders, and in that regard, we routinely face anti-American activities.

International terrorists have attacked us on American soil, and overseas as well. Attacks against our nationals, the military, governmental organizations and structures, and American business interests abroad are becoming all too common. As an example, anti-American attacks for the year 2000 worldwide against U.S. targets included a total of 206 incidents. Of that total, 178 were against U.S. business locations and interests.[20] The trend in recent years is that terrorist acts are less frequent but much more lethal.

Moreover, there is a growing fear of domestic terrorists such as paramilitary militias and millenarian cults that perform criminal acts without clear political gain or intent in cultivating their own beliefs or revenge. Such antisocial convictions include racial, religious, and cultural issues.

The intent of the criminal act, along with the suicidal nature of the terrorist in giving up life for his or her "cause" can only give rise to our need for awareness and security protection. Given the nature and reasoning for such an act, we can be sure that other attacks will continue to occur.

As we have noted, the favorite tool of the terrorist is the bomb, and the justification given for such an act is usually a religious, social, cultural (ethnic), or political agenda. Criminal extortion cannot be ruled out as a factor, but the criminal activity in these instances is for some gain, monetary or otherwise, rather than to punish or validate a point of view.

Consider also that the media are used very effectively by the terrorist. By killing, kidnapping, assassination, and bombing, the terrorist creates a news event, and the media are compelled to report such events. Reporting the acts and events can be said to be in the public's interest and the duty of a free press, but to acknowledge and project the terrorists' beliefs can be considered as propaganda that can only help in broadcasting their agenda. News editors sometimes overlook this dichotomy.

Assessment

The devices or agents described in this section have been classified and considered as weapons of mass destruction (WMD). Of the three described in Class One, the most difficult to manufacture, and of course the most lethal, would be the nuclear weapon. Unless the terrorist is involved with a country or political organization that has the ability to manufacture a nuclear device, mechanical bombs and chemical or biological methods are most likely to be used.

Following nuclear devices, of the two other agents, biological weapons are more lethal than chemical weapons. The Defense Department has stated that, regarding biological and chemical employment, "the low technology required lends itself to prolific and even potential terrorist use." These agents are more easily produced and dispersed and, therefore, are probably the terrorist's first choice. Based on present technologies, the employment of chemical weapons can cause the death of tens of thousands of people, and the use of biological weapons could kill hundreds of thousands of people. Readily available commercial materials and technologies found at pharmaceutical and chemical plants and makeshift laboratories, can produce these agents with ease.[21]

The more sophisticated devices for biological or chemical dispersal would require the skills of scientists and engineers, and the greater the cause, the greater is the chance of recruiting confederates with that expertise. Small quantities can cause extreme devastation and can be transported and smuggled easily.

Depending on the nature of the implement needed to spread the particular agent, we must consider that contaminants can be entered into the air that we breathe, the food we eat, or the municipal water supply that services what we drink.

The type of contaminant and the extent of the exposure may cause the effects of both agents to be immediate or delayed, in that it may take hours or days for the victims of the exposure to exhibit symptoms.

Chapter 9

Bombs and Bomb Threats

Although threats will be made and bombs may be placed by deranged and disgruntled persons, we are most concerned during our present time period with the terrorist act. With that in mind, we must consider the following factors, which could relate to any bomb threat or the activation of such a device, including the actions of a terrorist.[22]

The Warning

During this period of our country's history, many terrorist organizations, both internal and external, that are dedicated to disrupting the American way of life. By causing death, serious injury, and property damage, where the destruction causes an economic loss to the business under attack, they cause fear and apprehension among the innocent public at large as we go about our daily routines. They do this with the terrorist's primary weapon of choice: the bomb.

Bombs can be disguised in many ways and can be designed to cause specific damage and injuries. They can be planted indiscriminately or in specific places, depending on the access that the perpetrator may have or the purpose of his or her intent. The terrorist will usually target buildings or facilities that contain large numbers of people. As we have recently seen in Europe and the Middle East, businesses that attract and are open to large groups of people are viable and accommodating targets. In the Middle East in particular, many people fear public areas or the use of public transportation. In our country, it was a matter of time before terrorists would turn their energies away from our military and government facilities and include the business establishment here at home. The World Trade Center tragedy of September 11, 2001, in New York City had a serious and debilitating effect on America's financial and economic stability, resulting in the deaths of thousands and innumerable injuries to many, the destruction of a major landmark, and a profound effect on the American psyche. Before 9/11, a terrorist act of such magnitude was almost unimaginable. Consider it a lesson learned; we must understand that such an act may take place again in the future, and we must be prepared for it.

Terrorists are not the only people who may use a bomb for a particular purpose. Criminals may use a bomb (or the threat of one) for extortion purposes, as revenge for a perceived wrong, or simply through a desire to disrupt normal activities. Unless the perpetrator is mentally unstable, the reasoning behind the bomb threat is not to injure or cause damage, but to elicit money or something of value, or to cause a business interruption. These are criminal acts of another nature.

- Bomb threats without demands are to be considered most serious, since the purpose is usually for political or social reasons.
- Bomb threats with demands are usually threats by extortion, most often for money.

The following subsections discuss the various ways in which a business establishment can become aware of a threat or the presence of a bomb on premises.

Bomb Threat without Demands

The language of the threat is concise and deliberate, short and to the point. The caller gives little warning before the bomb will explode, and rarely is a time given as to when the bomb is set to explode. The threat may have been made to a third party: the police, the news media, or a telephone operator. As soon as the statement about the bomb is made, the perpetrator hangs up and answers no questions.

Under these circumstances, the threat may be considered genuine. Call for police assistance immediately. Whether a time of activation is given or not, the decision to evacuate the building must be made by management or as directed in the Emergency Procedure Plan.

Some of the reasons that the threat may be considered a hoax are that the caller gives more than a one-hour warning; the caller uses slurred speech and may be intoxicated; or the threat is made in a joking manner or is made in a fit of anger, where loud, abusive, or foul language is used. Management may wish to wait for the police to arrive before a determination is made for evacuation. This would strictly be a judgment call, one that many managers would not wish to make because it would delay the response to a potential threat.

A true terrorist will not give a telephone notification or any form of notification concerning a bomb that is about to explode. There will be no warning. In some instances, a group or organization will lay claim to the explosion after the occurrence. But in many cases, we can only assume who may have committed the act, or identification of the perpetrators is made only after an investigation.

Bomb Threat with Demands

This is the bomb threat by extortion, most often for money. A building manager or someone such as a company administrator will usually receive the call so that the caller will be able to speak to someone in authority and can expect directions to be carried out. The person receiving this type of call should attempt to delay the delivery for more than an hour, advising that only certain people are able to conduct a transaction of this type, and it will take that long to set up delivery of such a sum of money. Advise the police immediately after the caller hangs up.

Remember that the extortionist will try and put pressure on the manager with regard to time, but if his threat is in fact carried out and a bomb does explode, he has failed to achieve his goal, providing somewhat of an advantage. Any information concerning an extortion of this type should be strictly confidential, and such information must be shared only with the police and those who must be involved.

Threats Received by Telephone

Generally, if a bomb threat is received, it will be by telephone. Any company telephone through which the general public can contact the company should be manned by personnel who have been trained to remain as calm as possible and to obtain as much information as practical from the caller. Every telephone that can receive incoming calls must have a copy of a *bomb threat checklist* handy so that, by following directions on this form, the person receiving the call can easily attempt to gain as much information from the caller as possible. (See Appendix A for a copy of this form.) The employee receiving the threat most probably will be the company switchboard operator, receptionist, security guard/receptionist, or a customer service representative. Such employees must be trained and directed to write on the checklist all information that can be elicited from the caller and whatever can be remembered after the caller hangs up. The assessment of what the caller said, the time the device will or may activate, the possible location of the device, how the caller acted (attitude, manner of speech), and various other indications may well determine how the loss-prevention unit, administration, and the police will react to the threat.

The employee receiving the bomb threat must attempt to determine from the caller as much information as possible regarding the following:

- When the bomb will explode?
- Where the bomb will explode?
- What type of bomb is it?
- How or why it will explode?
- Why is the caller doing this?
- Who is the caller and whom does the caller represent?

Upon completion of the call, the employee should continue to complete and enter as much information on the ***bomb threat checklist*** as can be remembered, and to follow the instructions concerning notifications.

Threats or Bombs Received by Mail or Messenger

Letter or package bombs are made to kill and maim when opened. Such devices can be received from terrorists, extortionists, or deranged or disgruntled persons. It is not uncommon for this type of hazard to occur, and therefore it should not be taken lightly. The employee who receives or handles incoming mail or packages should be aware of the following information.

If a letter or package is received and the mail handler or the recipient has the slightest suspicion that it may contain a bomb, the loss-prevention department should be immediately notified. Once the suspicions have been confirmed, the police should be called. If there is any doubt, do not open the letter or package. Instead, it should be placed and isolated in a locked room away from windows and thin-partitioned walls to await the arrival of the police.

Listed below are a number of indications that should alert the mail handler or receiver that the letter or package might be considered suspicious and warrants further inspection:

- Excessive postage; there appears to be too many stamps for the weight of the envelope or package.
- Foreign mail, or the amount or type of postage may not agree with the type or class of package mail.
- A package that has excessive wrapping or is very poorly wrapped.
- Excessive material that secures the envelope or package, such as masking tape, duct tape, heavy string, etc.
- The envelope or package feels overly heavy for its size.
- Uneven weight distribution within the package, or the contents appear to be rigid within a flexible envelope.

- The envelope is uneven or the package is lopsided.
- As to the address or other wordage:
  - The handwriting or typing may be illegible.
  - The title, name, or common words may be misspelled.
  - The article may be wrongly addressed or have an incomplete correct address.
  - No return address is noted, or the return address is from an unknown or suspicious source.
  - The letter or package contains special markings, unusual endorsements, or restrictions such as "Fragile," "Handle with care," "Private," "Confidential," "Personal," "Addressee's eyes only," "Please deliver directly to ...," "Urgent," or "Rush."
- Greasy areas or grease/oil marks or discolorations are on the envelope or wrappings.
- The envelope or package has protruding wires or tinfoil.
- A strange odor or an odor of almonds or marzipan emanates from the envelope or package.
- There may be some type of resistance upon opening.
- The letter or package may have been delivered by hand from an unknown source and, if so, may be a timed device.

Written bomb threats received by messenger or by mail often provide excellent evidence for the police. Threats of this type are usually made for the purposes of extortion and imply that, if the demand is not complied with, an act of terror, most likely a bomb, will befall the company in question. Once a written threat is recognized as such, avoid any further or needless handling so as to preserve fingerprints, handwriting, postmarks, and any other markings on the letter or envelope for forensic examination. The loss-prevention unit should notify the police immediately upon discovery.

Finding a Suspicious Object without Warning

The security officer should realize that not all bomb devices placed in a business facility will come with a warning, particularly if it is the work of a terrorist. It is essential, therefore, that the security officer be constantly vigilant of any suspicious object or package noticed during officer's routine inspection or foot patrol of the facility. Basically, he or she should be concerned with anything that

- Should not be where it is found
- Cannot be accounted for
- Appears out of place

Add to this list what many law enforcement professionals consider to be an important investigative tool—"gut feelings" as to what you see, feel, or perceive.

The discovery or identification of such a suspicious package may also occur during a search by the police for a reported bomb on premises, and the same concerns must be considered.

In essence, the security officer's determination that an object or package is suspicious will be based on his or her knowledge and the layout of the facility, what could or should not be out of place, and a "gut feeling." Because the security officer is employed in a facility that is to be searched, he should have key access to all spaces and should be particularly knowledgeable of all spaces within the building where a search has to be made.

If a suspicious object or package is found under either of these circumstances, it should not be handled by anyone. The police are to be notified immediately.

Building Explosion

An explosion of some significance in or near a building can cause a partial or total building collapse. During such an event, fires will break out, thereby causing further destruction, injury, and death. Employees who are working in a building when an explosion occurs should be trained in the following procedures. In addition, employees who will be working in a multilevel building should follow the evacuation procedures as enumerated in Appendix A.

1. Review emergency evacuation procedures. Know where all emergency exits are located in and near your workstation.
2. Building maintenance should maintain emergency lighting according to municipal code.
3. Keep and maintain fire extinguishers in good working order as per municipal code. Know where fire extinguishers are located or be able to identify signage indicating the location of a fire extinguisher.
4. Learn basic first aid and CPR (cardiopulmonary resuscitation). Contact the local Red Cross chapter or private instructors for training.

Building owners and loss-prevention or safety managers should keep and maintain the following safety items in designated places (on each floor if the building is multilevel), particularly in the fire command station described in Chapter 18:

Portable hand-held radios with extra batteries
Several flashlights with extra batteries
First-aid kits with manuals
A sufficient number of hard hats for emergency responders

The necessary and appropriate tools that may be needed to clear debris or blocked exits
Tapes and roping to seal off or secure a dangerous area

See also "Protective and Emergency Equipment" on page 16.

Threat Analysis

For the decision-making process, an assessment of preparedness and potential targets or occurrences must include the following concerning a bomb threat or a possible bomb on premises.

- Attention should be given to those businesses that have international holdings or do business internationally. Consider that the larger or more well-known and established the company name, the vulnerability will rise accordingly. Also included are major utilities; schools and universities; research laboratories and abortion clinics; any federal, state, or local government building or structure; and any large structure, facility, or conveyance that will contain a large number of people in close contact with each other.
- During all bombing detonations, local, state, and federal public safety personnel and equipment will arrive to assist in control of the actual occurrence and the investigation thereafter.
- Generally, most bomb threats and warnings are delivered by telephone. The purpose of a bomb threat (or its aftereffect) is to cause panic, fear, and apprehension, and the intent of a bomb explosion (the specific event) is to enhance a specific objective that may be political, religious, ethnic, or cultural. And therefore, the definitive purpose is to create not only anxiety, but death, injury, and destruction.

Certain actions must be considered once a bomb threat is received:

1. The caller must not be placed on hold.
2. The receiver of the telephone call must compile a bomb report checklist during and after the call that notes all pertinent information. If the company has the ability, the call should be recorded.
3. All bomb threats must be considered as serious, with immediate notification to public safety authorities.
4. The company may react to a threat as follows:
   - Specific threat
     - The caller has definite knowledge of or believes that a bomb has in fact been placed in or around the facility in question, and wishes to minimize injury and damage. This caller may be the person who has

    placed the device or is someone who is aware that there is in fact a planted device.
  - The usual information given includes some or minimal facts about the bomb, where it might be located, the possible time of detonation, and who is responsible or why the bomb was placed.
- Nonspecific
  - The caller simply states that there is a bomb on premises. He or she gives no specific details as to the type of bomb, where it has been placed, who placed the bomb, and for what purpose. Usually the call will be short and to the point.
  - The call is a hoax and the caller wants to create as much disruption of the business operation as possible with an atmosphere of fear, panic, and apprehension.

Evacuations

One must consider that, in any evacuation scenario, people leaving the premises may become injured because of the rush to leave, confusion, apprehension, or panic. Under these circumstances, there is a probability that a minor or serious injury can take place, whether or not there is an actual emergency. Along with that, the likelihood of civil litigation because of an injury, subsequent emotional stress, or agitation cannot be overlooked.

An evacuation is a serious consequence that, once put into effect, cannot be canceled without more confusion. Consequently, it is most important that all employees be instructed and trained in how to react during such an event. Once the employee is made aware, by coded announcement or otherwise, of a possible emergency situation and the possibility of evacuation, that employee should adhere to the following guidelines:

- Become aware of conditions at your location, including where all fire extinguishers and fire exits are located. This should include a fast approximate headcount of customers, visitors, or employees in that department or area of concern.
- Once the evacuation announcement is made, direct all nonemployees to the closest fire exit and ask that they remove themselves away from the building. Evacuation procedures will be conducive to communicating the seriousness of the emergency at the facility in question (hospitals, nursing homes, correctional facilities, and other institutions).
- Do not panic—otherwise other coworkers and other occupants will pick up your demeanor. Attempt to act in a professional manner.
- Do not have anyone use any elevator. Have them use the closest stairway or fire exit to the outside.

- Before leaving, the last employee should make sure that all coworkers and nonemployees have left their area.
- If a visitor or customer refuses to leave or act at the employee's direction, the loss-prevention department should be advised immediately for its assistance.
- Once out of the building, the employee should meet and queue up with other coworkers from the same department at a prearranged location so that their manager or supervisor can take a head count. Anyone not accounted for should be made known to the loss-prevention department.
- To enable the management to take a correct head count of all personnel, the receptionist/switchboard operator or other designated person shall leave the building with the register or roster containing all work schedules, the visitor log, and a bullhorn. These items should be handed over to the individual intended to determine that all persons are safely evacuated.

Evacuation drills for training purposes should take place for employees at least twice a year, preferably every three months. These drills should be conducted before opening for business so as not to cause trepidation to others that may be on the premises during the routine business operation. Some institutions such as schools should hold drills routinely while school is in session. For safety, insurance, and liability concerns, an adequate record of the drills should be compiled and maintained for future reference.

There must be written policy as to whose order and under what conditions that an evacuation of customers, employees, students, visitors, and other occupants will take place.

Evacuation of the Premises

No matter what company management believes as to the seriousness of the threat, an evacuation of the premises of all occupants may still take place. The local law enforcement authorities may require an evacuation based on the threat received, on other local occurrences, or on their own intelligence information. Whether an evacuation takes place or not, a search of the premises will be made by security and loss-prevention personnel with the assistance of the police.

No Evacuation of the Premises

Realizing that very few bomb threats are authentic, a decision to take no action as to an evacuation may be considered. Company or facility management may believe that the threat is bogus and of no consequence. In addition, once the police are on board—and based upon their intelligence, information, and the facts given to them at the scene—the police may leave the decision to evacuate up to the management of the facility.

In the event that *no evacuation* will take place, the company or the manager of the facility must also realize that—if staff members are mistaken in not evacuating the facility, and if an incident does take place in which damage and injuries do occur, no matter how minimal or the cause—it will expose the company to civil and criminal litigation.

Claimed Responsibility for the Detonation of the Device

If there is a communication by the bomber or by a confederate regarding the activation of a bomb made prior to or after the activation of the device, we can assume that a reason will be given for the act. However, in an occurrence initiated by a terrorist or a terrorist organization, if, in fact, some notification is made, it will be made after the device explodes so as to advise why it was done and who is responsible for the act. This type of an occurrence most often takes place in a large public building or area, or a highly notable or identifiable location. As previously noted, it is usually intended to make a political or religious statement, and because of the resulting death, injury, and destruction, such acts will cause future fear and anxiety among the populace.

Employee Notification

All employees must be trained in the following procedures so that they may be aware and act accordingly in the event of a bomb threat.

The Initial Warning

Once the threat has been received, the following acceptable notification can be made via the public address system:

This is a (name of company) time check. The time is now…

In this manner, upon hearing this announcement, all employees will become aware that a bomb threat has been received, and all radios, pagers, and cordless phones are to be turned off. All assigned employees (including the fire brigade) are to report to the centralized location as directed in the Emergency Procedure Plan for bomb threat procedures.

The Evacuation

Evacuation will take place as per the Emergency Procedure Plan or the written bomb threat procedures.

Table 9.1 ATF Vehicle Bomb Explosion Hazard and Evacuation Distance Tables. Publication ATF15400.1

Vehicle Description	*Maximum Explosive Capacity*	*Lethal Air-Blast Range*	*Minimum Evacuation Distance*[a]	*Falling-Glass Hazard*[b]
Compact sedan	500 pounds 227 kilos (in trunk)	100 feet 30 meters (in trunk)	1,500 feet 457 meters	1,250 feet 381 meters
Full-size sedan	1,000 pounds 455 kilos (in trunk)	125 feet 38 meters (in trunk)	1,750 feet 534 meters	1,750 feet 534 meters
Passenger van or cargo van	4,000 pounds 1,818 kilos	200 feet 61 meters	2,750 feet 838 meters	2,750 feet 838 meters
Small box van (14-ft. box)	10,000 pounds 4,545 kilos	300 feet 91 meters	3,750 feet 1,143 meters	3,750 feet 1,143 meters
Large box van or water/fuel tank truck	30,000 pounds 13,636 kilos	450 feet 137 meters	6,500 feet 1,982 meters	6,500 feet 1,982 meters
Semitrailer	60,000 pounds 27,273 kilos	600 feet 183 meters	7,000 feet 2,134 meters	7,000 feet 2,134 meters

Source: Data acquired from the Bureau of Alcohol, Tobacco, Firearms and Explosives, a federal law enforcement authority now part of the Department of Homeland Security.

Note: Hazard ranges are based on open, level terrain; metric equivalent values are mathematically calculated; explosion confined within a structure may cause structural collapse and building debris hazards; additional hazards include vehicle debris, body parts, etc.

[a] Minimum evacuation distance is the range at which a life-threatening injury from blast or fragment hazards is likely; however, non-life-threatening injury or temporary hearing loss may also occur. Minimum evacuation distance may be less when explosion is confined within a structure.

[b] Falling-glass hazard range is dependent on the line of sight from explosion source to window. Hazard is from falling shards of broken glass.

See Appendix A for guidelines that any business enterprise can follow to evaluate the credibility of a bomb threat as well as procedures used to search for a bomb.

Precautionary Procedure

During a bomb squad training session, the question of who will participate in the search will arise. Police officers and loss prevention, security, and safety personnel are obligated by their job description to assist in a bomb search.
However, there may be other employees at the facility who possess and have key access to all spaces that require a search and these employees may be asked to assist in the search. These employees may involve managers, maintenance and/or engineering personnel.

It has been recognized that these employees may feel that actions or behaviors of this type do not include them as "emergency providers," particularly in a bomb search. A claim can be made that they are not police officers or security and safety personnel and thus their job descriptions do not entail assuming responsibilities or functions that might place them in dangerous positions.

Be aware that it has been held that anyone has a right to refuse to put himself or herself in possible danger. If a police officer refuses to act on a direct order from a superior, summary action will be forthcoming. So, too, security or safety officers face similar circumstances if they refuse to assist in a search. But no action can or will be taken against any other employee who refuses to participation. Their involvement must be completely voluntary and they should be listed with all other participating employees in the Incident Report for accountability and liability purposes.

Chapter 10

Other Serious Emergencies That Can Become Disasters

Sabotage

Sabotage may be defined as the intentional damage of property, thereby destroying, obstructing, or hindering the productivity of the normal business function. Sabotage also includes the use of treachery and subversive tactics to cause damage or to disable equipment or property of a business or government agency.

This act may be caused internally or externally by a person or group. It can include arson fire; the destruction or obstruction of telephone, gas, and electric lines; and the intentional vandalism of property in an attempt to damage, slow down, or shut down the viability of a business or government entity. Additionally, sabotage may include libel, slander, or the poisoning of a business product in order to harm the reputation or good will of a business enterprise.

All acts of this type should be reported to the police. Any act of sabotage committed by an employee of the business must be dealt with severely, in that if an employee is found to be involved in such an act, company policy will dictate arrest and prosecution to the full extent of the law.

See also Chapter 11, "Computer Theft and Sabotage."

Industrial Espionage

The risk of a company becoming a victim of information theft has increased in recent years with the advance of new technologies such as cell phones, cordless phones, faxes, computers, the Internet, machine copiers, microrecorders, and copiers. Industrial espionage is defined as the act of surveillance, infiltration, or spying on the activities of a business or government agency in order to steal information or something of value for oneself or another, such as a competitor or foreign government. Espionage includes the stealing of proprietary information that is considered to be of great value to the owner, where such a loss may cause a negative effect on a business or the safety of a nation. In essence, anyone or any company possessing information that is of value to another is at risk.

This information will include:

- Business plans, records, and programs
- Formulas, designs, prototypes, and physical devices
- Computer records, programs, and computer information
- Secret or technical information and formulas as well as prototype devices or objects
- Personnel information
- Proprietary lists of clients and customer databases

See "Intellectual Property: Intrusion and Protection" later in this chapter.

The Espionage Act of 1996[23] notes that to obtain any type of trade secret via unauthorized means of any type, where the release of such trade secret would cause economic harm to its rightful owner, is a federal crime. "Unauthorized means" include actions by undercover or nefarious employees occupying key positions of trust.

In 1995, The American Society for Industrial Security (ASIS) conducted a "Trends in Intellectual Property Loss Survey," in which 47% of the responding companies admitted to being victims of information theft within the last 12 months. Moreover, 27% acknowledged that they were unsure whether they had suffered a loss by such means. In 1998, an ASIS survey noted that U.S. business losses due to information theft were estimated at close to $300 billion.

Competitive Intelligence

Competitive intelligence is defined as the process of monitoring the competitive environment, company profiles, and research and development. It enables administrators to make informed decisions regarding marketing, R&D, investing tactics, and business strategies. It is considered to be a legal and ethical collection of information and analysis by one business against a competitor. It is not considered espionage.[24]

For more information, see "Competitive Intelligence: Company Profiles" in Appendix A.

Intellectual Property: Intrusion and Protection

Defined

Intellectual property can include a particular manufacturing process, an invention, a chemical or engineering formula for a new prototype or product, patent information, literary and artistic works, symbols, trademarks, brand names, and images. Proprietary information may be more of a loss to certain companies than others. As an example, pharmaceutical enterprises regard their research, development, and sales information as the most important physical assets, which must be considered the most important elements to be protected.

In regard to the various threats and vulnerabilities discussed earlier, there are four types of intellectual property that can be attributed to a company reversal if an infringement occurs. Any transgression will constitute a crime, and the individuals or companies committing such an act can be prosecuted.

Patent: The national or international registration of an invention or other tangible items, giving complete right to the patent holder concerning manufacturing and marketing rights. Patents last for 20 years and can be renewed. If not renewed, a patent is considered as expired, and the company loses the right to keep others from manufacturing and marketing the product.

Trademark: A name, phrase, sound, symbol, color, or corporate logo that is associated with a company or product that, once marketed and widely accepted, connects that trademark with the product without doubt. This trademark may also be called a service mark, and the legal protection is equal. Once registered with the Patent and Trademark Office, protection from competitors will last for 10 years and can be renewed. The symbols ®, ™, or "Reg. U.S. Pat. Off." must be placed next to the trademark.

Copyright: Copyright protects artistic expressions such as novels, textbooks, manuals, movies, poems, songs, and music that are "fixed in a tangible medium of expression." Essentially, it protects newly creative work placed in written form. Although it is not necessary to solicit a copyright, doing so gives protection to the author or developer if any infringement occurs. If the copyright is sought and granted, the symbol © along with the date of publication should be noted on the publication. A copyright lasts for the life of the author plus 50 years.

Trade secret: Any formula, recipe, pattern, device, or collection of data that gives an advantage of one company over another may be considered a trade secret. To prove that the item or information is proprietary, appropriate measures must be shown to protect the secret such as restricting knowledge of the secret to a select few. Moreover, it must be established that the information adds to the value of the business and should thus be kept a secret and that its loss would do great harm to the company. Any infringement of such a trade secret could cause the owner to seek damages.

Vulnerabilities

Be aware that the enemy may lie within…

Obstacles to maintaining a protected environment for intellectual property are as follows:

- Failure to encrypt sensitive computer files.
- Failure to shred or burn *all* sensitive papers and documents.
- Failure to conduct in-depth background checks in hiring personnel for sensitive positions and to maintain constant supervision and quality facilitation thereafter.
- The company executive on the road with a laptop computer containing highly classified or secret company information, including his or her company's latest and most vital activities, may leave a laptop unprotected in the hotel room. A spy entering this room can easily copy data from the hard drive. CDs or floppy disks, oftentimes located nearby in the room, may be available and given enough time, all data can be compromised. Encryption technology, which is available, can protect laptop data.
- Managers, administrators, or executives who have access to company reports, statistics, or important data can sell them to the highest bidder, or they could be compromised by blackmail and coerced into providing company information.
- Personnel who are in-house or contracted programmers could be unscrupulous, larcenous, or disgruntled and vulnerable to nefarious schemes. They could plant a "back door" into the company computer system so that repeated access can be made. For example, conducting money transfers to their benefit or complete sabotage of the whole system, which could seriously affect the company's operations and cause financial calamity.
- Foreign governments have been known to "bug" first-class seats on aircraft and hotel rooms frequented by executives.
- Computer data is much more easily obtained than information that is stored using the older method of paper reports.

Protection from Loss

Today, one of the more insidious exploits are achieved by criminals who commit copyright and trademark fraud by copying music or movies on CDs or DVDs, or producing look-alikes ("knockoffs") such as luggage, clothing, handbags, and jewelry. We can assume that the more expensive the item or merchandise, the more someone will attempt to copy it and offer it for sale to the unknowing customer at discounted prices. Not only is there a loss of earnings and good will (eventual discovery of poor workmanship and material) for the manufacturer, but also for the deceived customer who believed that the item was in fact genuine. Investigation of this type of crime would include the private investigator and a knowledgeable company employee who can positively identify recovered faux merchandise as bogus. Further, the use of a sting operation would be the most incontestable method for conviction and closing down the operation.

Nevertheless, the protection of intellectual property must customarily fall on the legal department of the business in question. Prosecution is difficult and recovering stolen information is even harder. The education of employees, particularly those who have access to intellectual property; the availability of tools and programs that track sensitive documents as to use, users, and location; good communication between the company's security and management groups regarding intelligence and perception of employees; and secure areas where such information is retained are a few methods of protecting this type of company asset.

Realize also that foreign businesses and governments are notorious for industrial espionage. European and Asian countries in particular have had reputations of this sort for years. In fact, considering that corporate and industrial espionage has become so extensive, the FBI and the CIA advised U.S. businesses in 2001 that these agencies were unable to protect American businesses and, in reality, they were on their own.

Most of the espionage committed within American companies is perpetrated by employees. Computer firewalls and assorted programs may prevent or at least hinder outsiders from gaining information, but the insider may have full access to intellectual property.

In addition to applying certain procedures for the management and protection of operations and devices of all intellectual property possessed by a business, the proprietary or private investigator must be aware of how an interloper—a professional industrial spy or a competitive intelligence expert (commonly called a *snoop*)—accomplishes his or her task.

Other than dumpster diving, intelligence is collected in bits and pieces, with the spy putting it all together for a positive result. Remember that this type of intelligence gathering is not only used by business competitors, but by foreign governments as well. Some examples of how information may be gathered are as follows:

- Private and personal information, particularly salacious details concerning an employee that, if made public, would place that person in some jeopardy unless the desired information is provided.
- A person pretending to be a student working on a research project, requesting meetings and information from personnel in a company's research and development (R&D) department.

- Persons who present themselves as potential customers at trade, industrial, or electronics shows where new or upcoming products are shown.
- Company employees who unknowingly or consciously post comments in Internet chat rooms and on Web sites.
- Lax security by company employees, particularly those in R&D departments.
- The shredding of sensitive papers, manuals, etc., instead of burning them in a secure facility by security personnel.
- Companies that employ or have publicity departments issuing press releases regarding new products or new patents.
- Complete listings of sensitive personnel, including biographical, professional, and academic information in company annual reports, brochures, etc.
- Failure to have confidentiality or nondisclosure agreements for sensitive personnel while in service and upon separation from the company.
- Suppliers boasting about sales on their Web sites.
- Job listings that describe R&D positions for certain companies, including job interviews where a competitor's employee or a hired investigator pretends to be a job seeker.
- Employees who have a propensity to talk, especially when a new product or idea is developed and, in their excitement fail to realize that they might be overheard.
- Public areas such as restaurants, bars, and airports located near the place of business where employees must know not to talk about sensitive business subjects.
- Public records such as those available from the Environmental Protection Agency or OSHA, whose regulations require that certain information be forthcoming on new or upcoming programs that are within the privy of that regulating agency.

Once loss prevention or an investigator has some knowledge of how intellectual property can be stolen and the safeguards that should be in place to prevent such thievery, it becomes easier to investigate a theft with a positive conclusion. However, as stated earlier, the prosecution of a competitor who has possession of the stolen information is difficult, and rarely is a recovery made that benefits the original owner.

See also Chapter 12, "Protection of Proprietary Information."

Chapter 11

Computer Theft and Sabotage

At this period in our history, with our ever-growing dependence on the computer and computer technology, consider that this branch of knowledge is a timely subject that affects our complete financial, economic, commercial, and industrial infrastructure—in essence, the business of America.

The protection of computers from unlawful activity by those who would gain access to personal and proprietary data requires both computer- and network-based measures. Firewalls and antimalware software, particularly for high-end applications, are considered an absolute necessity in order to enhance computer security. Programs such as intrusion detection systems (IDS) and intrusion prevention systems (IPS) identify and impede or block threats. At this time, these systems can only detect approximately 70% of illegal intrusions and therefore are only partially effective in reducing the potential of computer penetration. However, at present these systems can and do implement a measure of acceptable application policies and usage, along with document control policies.

After a crime or penetration is discovered or there is a probability of one, the entire computer workstation must be secured to protect and maintain the integrity of the scene. The business or institution may be inclined to keep the problem confidentially among a few employees or managers until there is a certainty that a crime was committed. At that time, along with company counsel, they may wish to advise the police and make a formal criminal complaint. Whether the police or the company will utilize and control the forensic experts will be based on an agreement between both parties. Most important, however, is that the police investigation be conducted under the authority and control of the police investigators and the local district attorney. During trial preparations, counsel for the defendant will definitely make use of its own computer experts to attack the way evidence was collected and preserved and its chain of custody. This is the reason why the primary emphasis must be on securing and protecting potential evidence under forensic guidelines.

In an effort to meet the growing onslaught of hacking and computer invasions of all types, there is one area that should be of some concern: the process of *penetration testing*. This is a procedure by which a company hires an outside computer expert or consultant who purposely attempts to hack into the corporate system to assess its vulnerabilities. However, the company and the consultant may open themselves to a legal liability if problems arise. This could include a predicament such as if the client's data is compromised or a third-party system is broken into or made vulnerable as part of or because of the test. The third party could bring a civil action against the consultant and the company that hired the consultant. Furthermore, an unscrupulous member of the penetration test might steal and resell corporate secrets. The specifics of the test should be documented in written agreements along with signed liability waivers prior to a test of this type.[25]

Emergence of Cyber Terrorism

There was a time that, for an auditor or investigator to discover a crime and track the criminal involved, the major investigative tool was to follow the paper trail. Today, with the emergence and evolution of the computer, the Internet, e-mail, financial and business software, and an assortment of electronic accessories entering the market daily, the processing of data has changed the business environment and the way that fraud examiners and investigators conduct white-collar crime investigations. Incriminating evidence leading to an indication of a crime and a trace of the perpetrator is no longer limited to a paper trail but can be found on hard drives, servers, and compact and floppy disks.

However, the information to be gleaned from the computer and its components and devices must be handled carefully and judiciously; otherwise the data sought could be lost or compromised. For this reason, forensic analysts who are experts in computers and computer science, whether in-house or contracted, should be the group to probe for the data sought.

Cyber Attacks

Cyber terrorism is distinct from computer crime. It is also different from economic sabotage and the art of "hacking." Essentially, it is the attack or the threat of attack against computers, networks, and the information and data contained therein for the purpose of causing serious consequences to business, industry, the government, and the public at large.

A prolific author on this subject is Professor Dorothy E. Denning at the Department of Defense Analysis, The Naval Postgraduate School, Monterey, California. She has been an active participant in all phases of computer science, working on cyber crime and cyber attacks for over 30 years. The following particulars were published under the title "Cyber Terrorism," and a summary of that article follows.

The following types of cyber terrorism were enumerated:

1. ***The destruction or corruption of data.*** Can be obtained by the use of viruses, malicious codes, malware, or any other means so as to destroy or corrupt files, changing or erasing data in databases or software systems.
2. ***The disabling of the computer system.*** A complete or partial disruption of the systems contained therein. This could include viruses, excessive spam, and service attacks to crash or disable the system.
3. ***Modifying computer output by penetration.*** Can be achieved by embedding a code or codes so that the computer will perform unauthorized functions then or at a later time.
4. ***Taking control of the computer system.*** Systems such as an air-traffic control system, a subway or other rail system, a 911 or any emergency telecommunication system, or a dam or flood control system, as examples, can cause severe consequences to life, limb, and property.
5. ***System penetration for theft.*** For the purpose of obtaining proprietary data and information, passwords, and other sensitive data.
6. ***Defacement of a Web site.*** By hacking into a Web site and changing the contents in order to spread misinformation, generate a loss of good will for the business, and to present a negative viewpoint for the public.
7. ***Illegal use of Web sites.*** By terrorists making use of chat rooms, encrypted e-mails, etc., so as to plan illegal activity, provide instructions and information to fellow activists, and recruit novices.

Obvious Systems for a Terrorist Attack

The systems that are of greatest concern are listed below. If terrorist attacks are carried out on these systems, they are to be considered severe aggressive acts against the nation as a whole.

1. Telecommunications: wireless, satellite, land lines, radio, and television
2. Banking and finance: ATM machines, Internet transactions, etc.
3. Transportation: air, rail, maritime, infrastructure
4. Oil and gas distribution and storage
5. Electronic power: nuclear, hydro, steam, etc.

6. Water supply and containment areas
7. Emergency services: police, fire, medical, etc.
8. Government services: particularly law enforcement, civil authority, and emergency health services

Protection Methods

In an effort to protect a proprietary computer system, the following procedures are offered:

1. Most importantly, having an on-site computer response team of individuals who have the training and expertise to deflect or intercede in any computer intrusion or crash
2. Having complete knowledge of what data and processes must be protected
3. Recognizing the threats and estimating the possible impact
4. Calculating the risks and deciding what level of risk can be acceptable
5. Developing and implementing countermeasures to reduce the risk as much as possible
6. Testing, regulating, and adjusting strategies to ensure security

Chapter 12

The Protection of Proprietary Information

The risk of having proprietary information compromised or jeopardized generally falls into four major categories.[26]

1. ***Verbal***: One of the greater risks to the loss of proprietary information is the employee who is privy to certain confidential data or information, and who unfortunately openly discusses such information in conversations he or she may consider private but in fact are in public forums where others can overhear them. Professional intelligence professionals seeking information from their employers' competitors will use every trick in the trade using the target as a new found friend or exploiting the naiveté of the subject to talk business. They are known to frequent local restaurants or hangouts, where the opportunity exists to overhear conversations around them. Moreover, employees are often the subject of telephone calls from individuals specifically trained to elicit information that could be revealed unknowingly.
2. ***Visual***: Critical information left on chalk—or whiteboards after planning or marketing sessions without giving any thought to who might walk in or come across these details jeopardizes the information. Critical data could include pricing information, financial risks, and negotiation points that would definitely afford the competitor inside intelligence that could help him. Creative observation by professionals can include varied means such as lip reading the competitors' employees at a luncheon or careful observation of staff actions, an increase in certain skilled personnel, and activity in and around

a competitor's facility. The more creative the professional, the more information can be had.

3. ***Physical***: Physical risks can be as simple as poor access control in and to the facility itself. Proprietary information can be easily had from company computers by expert hackers, who may not need to be on site to take the information. Professionals or dishonest employees need only to transfer data to CDs, floppy disks, or hand-held flash drives and walk off the premises, with company security unaware of the loss until it's too late. In addition, the careless disposal of drafts, memos, and the like in trash containers without shredding or the controls leading to the burning of this material can be of great risk. "Dumpster diving" or "garbage hunting" for any type of company data or information is a common tactic today. Moreover, the theft of laptop computers left unsecured in hotel rooms or carelessly left or placed in transport conveyances can have the confidential data extracted, even if encryption software is present.
4. ***Electronic***: Today, company data resides in electronic bits and bytes within the company's networked computer system. Without any doubt, it is increasing daily. However, the electronic environment contains many exposures, particularly so if the computer network is connected to the outside world. Since most company networks are exposed in this manner, these systems can be hacked easily from anywhere in the world. Safeguards exist and are upgraded continually, but expert hackers—with time on their side—can override these defenses. No perfect safeguard exists. Any hacker can make use of sophisticated programming tools to analyze a system for vulnerabilities.

However, most of the problems that exist today are suffered by companies that do not properly safeguard their systems from known loopholes and monitor for signs of unauthorized access. Additionally, communication via e-mails or the "chat rooms" that are used by employees engaged with other employees or friends—or even stock message boards for communication with the general public—may contain proprietary information that can circumvent a company's physical protections. Such breaches can occur through the use of both company and personal computers. Employees who are educated in the pitfalls of computer use will, sooner or later, become negligent. As a company hires service providers and makes use of consultants who need access to the Internet as part of their employment, the company's computers are at increasingly greater risk.

IT Security

Security of information technology (IT) is a challenge, as the pool of employees accessing the corporate computer network in any organization changes constantly, either through promotion, layoffs, or hiring and firing, thereby increasing security transgressions. Moreover, IT security threats from outside the corporate network evolve continually; therefore new methods to obstruct these threats must also evolve.

Computers have been used increasingly over the last two decades, and they are now an integral part of every business. Today's computers are used for management and operational functions, stock control, financial and payroll records, and all types of business records, plans, prototypes, formulas, and projections, which may or may not be covered by copyright or trademark. However, this information can be exposed to internal and external observation, sabotage, and espionage. Files and information can be taken off premises surreptitiously via small floppy disks and flash drives, or they can easily be sent electronically to another computer outside the company within seconds. Proper programming instructions must be used to protect information in the computer; otherwise, access or loss may never be detected. This includes intrusion from outside computers to elicit information or to create problems that can sabotage the business.

Consider the possibility that a person bent on computer sabotage or espionage might be a trusted employee who, as part of his or her duties, has access to all computer programs or files. Then there is the additional danger of an external hacker entering a company's computer system, particularly if the computer is connected to online servers or other sites off premises. Loss of business information and files along with destructive viruses have caused severe monetary losses and business disruptions. Today the computer is an essential part of business operations, and any intrusion or interference can cause extreme harm and business confusion. The company's resources and risk profile must be considered. In this regard, and depending on how important the computer function is to a business, the establishment of a ***computer incident response team*** (CIRT) may be considered critical. This response team must have the authority and support of the company administration to act immediately and effectively once a security breach occurs or a problem is identified. Written policy should delineate what a computer security emergency is, the team members assigned to CIRT, the team members who will respond to a particular problem, and the steps involved in attempting to correct the problem. In effect, this team should include the necessary in-house employees with the complementary skills required to immediately control the breach and begin system recovery. The company would be wise to also have outside expert professionals or technicians on call who will be familiar with the computer systems and who will respond to a computer emergency.

Another consideration that must be addressed is the theft of information from small hand-held computer devices. A variety of mobile instruments (removable cards, compact flash drives, etc.) can be connected to corporate networks for the storage, retrieval, and transfer of confidential data. Such data can be easily stolen or, in the case of sabotage, lost. The intruder can secretly and easily remove hand-held devices of this type, along with the stolen data, from a facility. These units can also be easily lost or stolen, and an adversary could use the data to compromise the interests of the original owner or the company.

The Four Deadly Security Sins[27]

Computer Insecurity

According to Scott Montgomery, global vice president for product management at Secure Computing,[28] if an organization relies on its staff to ensure that its network is secure, it already has a problem, as employees are not infallible in their use of company computers, and one slip would be all it takes for a cyber criminal to intrude or attack the company's computer system. He notes that the end users are typically the least educated in the use of proper corporate security practices, and they are the most prone in not adhering to the company's computer security policy. He highlighted the four most damaging security habits that are commonplace in the United States and around the world, and emphasized the need for IT administrators to closely control and manage these systems with meticulous inspection procedures. The four damaging security sins are as follows:

1. ***Using fixed passwords***: The Sans Institute[29] has identified passwords as one of the top 10 most damaging security practices. Unlike token-generated or one-time passwords, Montgomery noted that fixed passwords generally do not change and that many users write down the password in or around their workspace so that it can be retrieved if forgotten. Moreover, a picture on a wall or desk, or a favorite item within the workspace, can offer a clue to a password. A cyber criminal will make use of this failing, and once the password is found, the criminal can log into the network and cannot be identified as an imposter. The tactic of password guessing is negated with the use of one-time passwords, which reduce unlawful entry and increase the company's security profile. Another way to prevent unauthorized use would be the addition of a hardware token for the practice of one-time-password entry.
2. ***Neglecting inbound threats***: This would include e-mail, the Web, and instant messaging. For instance, when an end user receives a spam message in the e-mail inbox, the battle has already been lost. The reasoning here is that administrators expect the user to do the right thing (delete the spam message without opening it), but users often do the exact opposite. Most employees are not aware of the great risk they bring to their network by opening the message. Montgomery noted that these areas are to be considered high risk, and IT administrators must strictly enforce correct procedures through training and the probability of punishment for the laggard.
3. ***Neglecting the fact that data traffic is a two-way street***: The organization's network must be kept as secure as possible, and IT administrators should

drive home the point that data traffic is bidirectional and emphasize the possibility of outbound data leakage. Companies often protect only the data that enters their network but hardly consider the data that is leaving their network. Such activity may be caused due to malicious or criminal intent, or it may simply be the result of actions by a poorly trained employee.

4. ***Neglecting the need for encryption***: Encryption of e-mail provides a level of protection. Without encryption, data sent via e-mail is open to the view of anyone in the public. If someone wishes to look and find entry, it can be done easily by hacking into any network.

Cyber Criminal and Terrorist Activity

Microsoft Internet Explorer® is the most popular browser used for Web surfing. Older versions of Internet Explorer contain multiple vulnerabilities. In addition, Internet Explorer has been leveraged to exploit vulnerabilities in other various operating systems, core Windows components such as HTML, Help, ActiveX controls, and the graphics rendering engine. Microsoft makes patches and updates available routinely to correct these problems.

In 2001, the National Infrastructure Protection Center (NIPC) released a document summarizing the 10 most critical Internet security vulnerabilities. Succeeding years have expanded the list to 20 impediments. NIPC, located in FBI headquarters in Washington, DC, is an interagency center and includes representatives from the FBI, DOD (Department of Defense), DHS (Department of Homeland Security), military services, the intelligence community, various federal agencies, local law enforcement agencies, and private industry.

NIPC operations include the following:

- The ***computer investigations and operation section*** (CIOS) This section is the operational and response arm of the center and addresses cyber terrorism, cyber crime, and computer-based foreign intelligence activities.
- The ***analysis warning section*** (AWS)
- The ***training, outreach and strategy section*** (TOSS)

In addition, the FBI has created an investigational system known as the National Infrastructure Protection and Computer Intrusion Program (NIPCIP) in all its field offices both nationally and internationally. Recently, NIPC has been fully integrated into the Information Analysis and Infrastructure Protection Directorate (IAIPD) of the Department of Homeland Security. Moreover, as part of the National Cyber Security System, the United States Computer Emergency Readiness Team (US-CERT), a partnership between the DHS National Cyber Security Division (NCSD) and the private sector, has been established to protect our nation's industrial Internet infrastructure.

Further information may be found at the National Infrastructure Coordinating Center (NICC), an extension of the Homeland Security Operations Center (www.dhs.gov), and the ANSIR program, is conducted by the FBI's field offices.

The Law

Under the Electronic Communications Privacy Act[30] (unlawful access to stored communications), federal law mandates that it is unlawful to:

- Access without authorization any facility through which an electronic communication service is provided
- Intentionally exceed an authorization to access that facility and thereby obtain, alter, or prevent authorized access to a wire or electronic communication while such is in electronic storage within the system.

Some problems that can arise are additional programming instructions entered by an employee to destroy data so as to hide a crime, issue false payroll checks to nonexistent employees, issue checks to fictitious invoices, or cause downtime by intentionally creating errors. More common is the unauthorized use of computer time to play games or to conduct personal business on the company computer system.

Most computer crimes can be avoided by proper planning tailored to a business prior to the installation of the computer system. If a company is not large enough to have a competent computer expert or programmer, then one should be contracted to provide security, safeguards, and immediate restoration on an as-needed basis.

Security Precautions

Wireless Transmission Technology

The network over which wireless transmissions travel is, in actuality, "thin air" and cannot be secured like the traditional coaxial or Ethernet cable systems. Because of this perception, many people believe that wireless technologies are highly vulnerable to attack by an outside force because conventional security precautions are useless. However, with proper authentication and encryption technology, wireless is equally as secure, if not more secure than, wired networking systems. Without a physical cable to splice into, the wireless signal is much more difficult to intercept.

Therefore, in choosing a wireless-system product, the user should verify that it at least contains standard encryption and authentication processes to ensure the highest level of security. It would do well for the loss-prevention professional to work closely with the facility's onboard computer incident response team (CIRT) or the company's contracted computer experts so as to understand the various wireless technologies that are available and how each may be used in the operation at hand.

Any business computer open to electronic commerce or connected via outside communications is vulnerable to intrusion, whether perpetrated by a terrorist, "cracker," "hacker," or simply someone bent on causing sabotage or espionage.

An attack by terrorists who would hack into the computer networks of governmental, institutional, business, and private systems is not only possible, but also probable. The intent would be to sabotage all systems that include land-line and cell telephones; business, government, and personal computers; fax; e-mail; television; radio; commercial airlines; military aircraft; banking and business enterprises; medical care; and delivery of consumer goods, all of which rely heavily on computers to function. A disaster of this magnitude would be equivalent to an outright terrorist nuclear attack, where chaos would be the result—in reality, an "electronic Pearl Harbor."[31]

We must also consider the probability that intrusion may also be an internal problem, such as a disgruntled or larcenous employee, or an ordinary system crash. The business establishment should recognize the need for quick response to computer security incidents or major system crashes in an effort to protect, or at least save, as much computer information as possible.

In relation to that, large business operations should have a group of computer-knowledgeable people on site, or at corporate headquarters, who can respond to any computer emergency. They must have complementary skills and be able to work independently, particularly in high-pressure incidents. This computer incident response team (CIRT) must have complete authority in incidents of this type, along with adequate funding and the complete support of the corporation. Moreover, outside professional consultants or contractors may also serve as associates to the team and be called upon as needed. In any event, some members of the team may not necessarily require computer expertise, as in the case where criminal intent is found or an employee is identified as the culprit. In such instances, the loss-prevention manager, human resource manager, and a company attorney may be necessary parts of that team.

If the team operates out of corporate headquarters servicing various sites elsewhere, their duties will usually be complementary to computers, such as programming, upgrading, developing policy and security, monitoring and enforcement, and correcting ongoing minor problems. A team that operates in-house may have other duties not necessarily wholly connected with computer operations. In any event, if a serious computer problem or incident does occur, members of both teams must be able to stop whatever they are doing and direct their full attention to assisting in system protection and recovery.

A response plan should be drafted to include the required notifications to senior management, clients, and other business partners. The assigned members of the CIRT and the relevant response procedures should be made part of the *Emergency Procedure Plan*. Therefore, pertinent to computer security, some or all of the following precautions that may be related to a business operation should be established:

- All managers and employees who will have access to computer files, particularly sensitive business files and records, should be thoroughly interviewed and investigated during the hiring process.
- Repair personnel must be screened prior to working on any company computer or peripheral equipment. This should also include internal telephone equipment and external telecommunication lines.
- Sensitive computer operations should be placed in a secured or isolated area where only those employees granted specific access are permitted.
- Some type of a security escort should accompany any repairman or visitor to this secured area.
- Backup files and records must be kept in case of system crashes, deliberate or not. For safety, they should be secured in a locked fireproof safe or cabinet. If necessary, a second or master set of backup files may be stored safely secured off premises.
- Distribution of printout sheets and disks should be available only to certain authorized people, and so recorded.
- Loss-prevention audits should include routine inspections in these areas.

States and the federal government are of the opinion that computer fraud, theft, and damage are a serious threat to business, to the consumer, and to national security. Legislation has been drafted in recent years to cover this ever-increasing problem. Because a business establishment makes use of the computer in many areas (business plans, financials, purchasing, stock control, record keeping, etc.), the loss-prevention officer must be aware of the probabilities that may occur in varied situations.

In that regard, state criminal codes may define offenses involving computers in various terms. This may include a person who may be considered guilty of a crime when he or she knowingly uses (or causes to be used) a computer or computer service without authorization when such a computer is equipped or programmed with a device or coding system designed to prevent the unauthorized use of said computer or computer system.

In addition, the various classes or types of computer crime could include trespass or tampering, duplication, and possession of computer-related material, any of which may be considered a misdemeanor or a felony, depending on the state law in which the security officer is employed. The security officer may wish to review those sections of the relevant state's criminal code for future reference.

Consider also that computer fraud and related activity in connection with computers covers the unauthorized entry and access of a computer that is determined by the U.S. government to be protected from disclosure. If a person obtains any information or causes the transmission of such information or a code to another computer or computer system, or causes damage to any information contained therein—programs, equipment, etc.—then that person is deemed to have committed a crime.[32]

Conclusion

All computers should have individual password protection, no matter what types of access or programs are available. A click of the "on" switch without further safeguards offers no protection whatsoever. According to many professionals in the computer security field, an employee's password can be found on that employee's desk. If the actual password is not readily available, then it is likely that the employee's workstation contains various clues as to how the password was derived. A picture of a wife or girlfriend could mean a password educed from a name, nickname, or birth date. A pet's name, a sports banner, or paraphernalia on a desk may also provide clues to a password.

Computers logged on and left unattended can compromise an entire network, and information can be extracted and placed on a compact disk or a flash drive and then be easily taken from the facility in a shirt pocket.

Finally, bear in mind that any information stored on the computer's hard drive is vulnerable to compromise, destruction, or modification. Although any information can be encrypted to prevent some of these problems, the information can still be destroyed maliciously. A company might wish to leave information open on a hard drive so that other computers on the network can access this information, and therein lies the problem—the whole network is then vulnerable. The only way to prevent this is to reduce the risk by placing data on CDs or disks offline and as backup copies, and then placing these records or files safely in storage secure from destruction, fire, and theft.

Note that the threats and perils of terrorism, sabotage, and espionage that can damage a private corporation may also take place against a government, a national agency, or a significant national entity. No computer operation is completely safe from intrusion. Remember that constant vigilance is the fundamental element in computer security.

SECTION 3

Chapter 13

Natural, Accidental, and Intentional Occurrences

Whether an occurrence is natural, accidental, or human-made, the effect on people and their way of life will depend on the severity of the event and the subsequent response to that event. Beyond the intended terrorist or criminal act, the inhabitants of our planet face a variety of natural and accidental disasters and catastrophes on a routine basis. Unlike bombs or the chemical, biological, or fire emergency circumstances included elsewhere in this book, the incidents described below have the potential to cause severe damage to our environment and infrastructure as well as resulting in an enormous number of deaths and injuries.

Human-Made Disasters and Catastrophes

In the book, *Catastrophe: Risk and Response*, author Richard Posner submits in some detail the disasters summarized below.[33] Some of these catastrophes occur routinely, and some are the result of humankind's continuing adverse effect on the environment. The importance of science—or the failure to recognize its importance—will have serious consequences on our planet.

Posner divides his thesis into natural or human-made disasters, the latter of which are further divided into three categories:

- ***Scientific accidents:*** Technology gone wrong; miscalculation or poor prototype agents or mechanisms

- ***Unintended human-made catastrophes***: Accidental production or release of new viruses and bacteria into the biosphere; poor or negligent oversight of nuclear, chemical, or power plants; miscalculation or illogical response to a nuclear threat
- ***Intentional catastrophes***: Terrorist or criminal acts that cause death and destruction

Haphazard and Unplanned Disasters

Science and industry are driving technology so fast that what is exceptional today will be out of date tomorrow.

As biotechnicians attempt to find cures for unrestrained viruses and bacteria, there is the great possibility of an accidental production of new and lethal germs. The chances of this type of accident is greater in the manufacture of a bioweapon, where caution is put aside for the haste of immediate production.[34] Moreover, scientific manipulation of subatomic particles is not so well understood by physicists that an end-of-the-world scenario can be dismissed as total fantasy. "Concern with possible catastrophic consequences of particle-accelerator experiments led the director of the Brookhaven National Laboratory to commission a risk assessment by a committee of physicists" to examine the problem.[35]

Consider also the ongoing technology and potential effects of laboratory accidents involving nanomachines, human genetic research, and genetically modified crops.[36] Although only partially revealed here, these possibilities are of great concern within the scientific establishment.

Chapter 14

Natural and Common Environmental Occurrences

Natural disasters come in many forms. Excessive rain, flooding, hurricanes, tornadoes, and earthquakes are only a few of the natural events that can damage property, disrupt business, and cause a loss of assets. Damage caused by these natural occurrences can affect water, gas, electric, and communication lines. Consider also the deaths and injuries along with property losses and the relocation of those inhabitants affected. Earthquakes cannot be predicted and can occur without warning, and volcanoes and tornadoes occur with little, if any, warning. These occcurrences will cause not only severe damage and injury but fear and panic.

How much damage has been caused, how much of a loss has been suffered, and the length of the business suspension will determine the reaction of employees and the security force. The role of the security officer is to prevent further damage to property and merchandise, to tally the cost of the damage so that some recovery can be made through insurance or write-offs, and to help in the restoration of the business. But most importantly, if the disaster takes place during business hours, the basic response by security and safety officers is the concern for the protection of all of the inhabitants in the facility at the time of the occurrence and immediately thereafter.

Some of the following events or occurrences can happen without warning or within minutes of the initial event, and some may occur over a period of hours or days.

Pandemic

The 1918–1919 flu pandemic was an unexpected lethal variant of the flu virus that can remind us what nature can do to a large population. The disease spread far and wide, crossing borders, and no person was isolated from the possibility of infection. Millions were infected and millions died. Because the flu virus has the ability to mutate, there is the possibility that an even more potent form than the 1918 outbreak could emerge in the future somewhere in the world.

Epidemic Diseases

The seriousness of the outbreak will depend on the type of disease, the target population, and the geographical area covered and whether it is caused by a terrorist or criminal action or by natural phenomena.

The HIV strain that causes AIDS is an ongoing pandemic affecting people worldwide. Science has made some progress in hindering the initial virus from elevating into AIDS in some patients, but since first discovered as a serious threat in the early 1980s, the pandemic is still not under control, and the threat appears to be escalating. This is mainly because, once a person has become infected, the virus can lay dormant in the host for years while still being highly contagious upon contact, thereby maximizing its spread.

Earthquakes and Volcanic Eruption

Earthquakes

The movement of the earth, in which property damage can be severe with great loss of life, occurs without warning. In reality, this activity alone causes little death and injury; most casualties will be caused by falling objects, structural failure, collapse, and fire. Aftershocks can cause further damage and injury and can interrupt recovery actions. Earthquakes can also cause an extensive loss of fire protection by disrupting the water supply. Because of damage to highways and roadways, emergency response and transport will also be affected to some degree.

Nuclear Plants and Seismic Events

Another area of concern is nuclear plants and their location near large populations and areas with potential seismic activity. At this time, with more and more countries initiating the nuclear option for energy, safety experts are taking a hard look at international standards, guides, codes, and regulations of a nuclear plant's ability to withstand seismic events. Authorities must take into account the site evaluation of a nuclear plant within areas of active earthquake movement or consider the possibility of such activity that would place a nuclear plant in a precarious position. These evaluations of seismic hazards and the seismic response of structures must be carefully considered before the creation of any nuclear site.

Realizing that security and safety must be specific to nuclear hazards, we must also identify ways to remove obstacles that impede the safe transport of radioactive material and radioactive waste management.

Tidal Wave (Tsunami)

Tsunamis are a matter of concern in seacoast areas in the vicinity of an earthquake. The turbulence created by these events is a serious threat as population growth encroaches onto seaside communities.

Volcanic Eruption

Unless the eruption is a major cataclysmic event in which hot molten lava will destroy everything in its path, concern should be directed toward ash and gases within pyroclastic flows (or clouds) at temperatures as high as 1200°C. This flow or cloud, which contains carbon and sulfur dioxide and molten ash too heavy to enter the atmosphere, will destroy everything in its path as it races away from the lava flow at speeds up to 100+ mph. Lighter ash and gas thrown from the mouth of the volcano can reach as high as the stratosphere, inundating cities and large geographic areas as the ash descends and covers everything. Note the explosion of Mount St. Helen's in Oregon, where tremendous amounts of ash covered large areas. In particular, the city of Portland, Oregon, was covered with several inches of ash, causing respiratory problems to its citizens, a loss of production, and disruption of transportation.

To date, science has been unable to predict earth movement or volcanic eruption that could have an adverse effect on a specific geographical area. Earthquakes occur almost daily somewhere and many cause monstrous and destructive effects to life and property. At present, America's west coast is vulnerable to earthquakes and volcanic activity.

See "Volcano and Earthquake History" in Appendix A.

Tornadoes

Tornadoes can cause extensive damage to property and loss of life, and they remain an ongoing threat, especially in the central United States. Predictability is inaccurate, and there is very little warning to the populace prior to an occurrence. Because an event of this type is so unpredictable and erratic, severe destruction, death, and injury in some areas will occur, with little or no effect on structures adjacent to the path of the tornado.

Hurricanes

Hurricanes can cause extensive damage to property and loss of life. They bring excessive rain, extreme winds, and flooding, especially in the coastal areas of the eastern, southeastern, and Gulf Coast states of the United States. Hurricanes are more predictable than many other natural occurrences. By observing the hurricane's origin and path, communities can take some precautions and prepare for the event.

Excessive Rain, Snow, and Blizzards

Excessive rain, snow, and blizzards can cause flash floods as well as landslides and mudslides. Rain is the major factor in the flash flooding and landslides that cause harm to life and property. A major precipitation event is generally predictable, particularly because of increasingly reliable long-range forecasts. Loss of life is usually minimal, but property damage and the transportation of people and goods can cause a cessation or slowdown in the activity of commerce, transit, transport, and personal movement.

Dam or levee failure can be caused by excessive rain, resulting in complete or partial inundation of small or large geographical areas and causing death, injury, and property destruction.

Coastal Flooding

The Pacific, Atlantic, and Gulf of Mexico coastal areas of the United States, along with bays, tidal rivers, and the Great Lakes, contain "watershed counties" and account for 15% of our land mass subject to flooding. According to the National Oceanic and Atmospheric Administration (NOAA) and the 2000 census, 48.9% of the U.S. population lives within 50 miles of a coastline, and these 673 coastal counties constitute only about 25% of the country's land mass.

See "Coastal Areas of the United States" in Appendix A for more information.

Global Warming

Atmospheric scientists believe that future summers will appear to be much hotter than originally predicted. There is consensus that, by the mid-2080s, daily highs will be 10°F warmer in the eastern United States, and there is a growing belief that we must take strong action now to curtail the carbon dioxide emissions that are the source of global warming.[37]

Global warming is considered to be a matter of high urgency by the scientific community because, as Earth's ever-growing population increases, the need for greater economic growth is multiplied. This economic growth is currently predicated on the use of carbon-based fossil fuels (production of carbon dioxide), and deforestation (absorption of carbon dioxide) in many parts of the world has become a cause and effect generally acknowledged by the scientific community. Scenarios that are affected by carbon dioxide and other greenhouse gases include the following:

- Concerns regarding the correct accumulation or dissipation of carbon dioxide, methane, and other greenhouse gases that make the Earth habitable because they do not block sunlight from reaching the Earth, but do reflect heat radiated skyward from the Earth's surface; lacking this delicate balance in our atmosphere, the Earth would become frigid or extremely hot, significantly affecting resources and all creatures living on the planet
- Higher temperatures, which will cause a reduction in agricultural production worldwide, the result of drought, erosion, and storms
- Increased flooding, coastal erosion, and a reduction of coastal land caused by melting ice caps
- Increased retention of water vapor in the atmosphere rather than returning to Earth as rain or snow, a consequence of warmer temperatures in the atmosphere, contributing to an increase in wildfires and groundwater contamination
- Increased severity of storms—hurricanes, tornadoes, rain, snow, and flooding—caused by climatic changes and subsequent significant changes in ocean temperatures and currents
- Excessive drought caused by lack of precipitation, irrigation, or waterways; the major cause of famine and damage to the soil and environment

Chapter 15

Accidental, Intentional, or Unintentional Acts

Wildfire

Wildfire is caused by natural forces (dry land, lightning strikes) or by humans in large tracts of woodlands, which encroach upon residential and business areas. These types of fires are much more prevalent in the western states. Depending on the severity and size of the blaze, the need for timely evacuation and protection of property may be minimal.

The prediction and forecasting methods by national and local authorities of climatic and natural occurrences, the professionally attended national overlooks by forest rangers, and the information they provide can be of great use to both public and private enterprises in the preparation of necessary precautions.

Fire

Fire is the number one cause of death and property destruction in the United States. Experts agree that children and the elderly die of fires more than any other age group, especially in a residential setting, due to a lack of safety knowledge and preventive devices. See Chapter 18, Fire Emergency, for a full description.

Arson: Intentionally set fire, leading to destruction of property, death, and injury.

Accidental fire: A natural or mechanical response; unintentional and incidental to life and/or property.

Either event can cause death, injury, and property damage, and could include wildfire or a structural fire—natural, intentional, or accidental. See Chapter 18, Fire Emergency, for further information.

Gas Leaks

Gas leaks must be considered serious and potentially dangerous because of the insidious nature of the substance. Basically, the reporting, response, and possible evacuation procedures would be similar to those used when responding to a fire emergency.

In addition to the request for help from the local gas company in response to a leak, the fire department should also be requested to your location in case of an explosion or fire if the leak is severe. The fire brigade and all security officers should also be trained in this type of response to avoid causing a more serious problem with the use of electrical contacts, sparks, cigarettes, flashlights, or mobile radios. Investigation of the leak source should include steps to prevent an explosion.

Investigate all reports of a possible gas leak or the smell of gas thoroughly, whether inside or outside the premises. Consider all gas leaks as a potential source for explosion and fire.

Threat of explosion: Causing death, injury, and property damage.
Noxious vapors: Causing illness and death.

Chemical Spills

A chemical substance can cause harmful physical effects to the public at large. Accidental exposures generally occur from leaks and spills, and depending on the type of chemical, may contaminate large geographical areas. Such events include contact at or within the workplace. Overturned rail or truck tankers containing chemical, toxic, or noxious substances are a serious threat to people and the environment and will require the presence of public-safety responders to control the incident.

Human Events

Riots/civil unrest: Includes strikes, sporting events, assassinations, shootings, and hostage situations.

Workplace violence/aberrant behavior: Includes violent, disgruntled, or abnormal persons with criminal or erratic demeanor, causing death, injury, and property damage. Also includes employee or personal confrontations, altercations, and serious assaults.

Medical emergencies: Accidental injuries and illness on or off the site, suffered by employees, nonemployees, students, customers, clients, visitors, etc.

Mass hysteria and anguish: A severe condition that can grow within the populace and can be caused by any one of the disasters or emergency conditions noted herein. Consider that—depending on the severity of the disaster and as the incident perseveres, along with little or slow response in aid—rational and emotional discord is amplified as every negative development is made known. This effect can be easily passed among groups of people; it can be contagious and its movement to other groups can easily occur.

Public Events

Transportation

Transportation incidents: Intentional or accidental occurrences to aircraft, airlines, rail, subway, maritime, and all other public conveyances for hire, etc., caused by various means, which may include terrorist, criminal, or accidental factors. It will involve death, injury, property damage, and the loss or breakdown of the transport infrastructure.

Public Health Concerns and Hazardous Materials

Transmission of bloodborne pathogens: Exposure to harmful bacteria and viruses.

Biological, chemical, and toxic exposure: Intentional or accidental contamination or hazardous spills or leaks.

Radiation release: By a nuclear plant or facility; accidental or intentional; also dirty bomb release.

Food poisoning: Accidental, negligent, or intentionally, causing a mild or serious illness.

Accidents to Property

Unintentional and unforeseen: Facility/auto/property damage.

Building/structure collapse: Damage, death, and/or injuries.

Specific Business Disruptions

Economic/financial collapse or interruption: Sabotage or espionage, or simply computer interference or crashes.

Sabotage of electronic and computerized programs and equipment: By terrorists, deranged or disgruntled employees, or those with criminal intent to commit extortion or fraud.

Strikes and picketing actions: Legal or illegal (walkout, wildcat strike); frustration among the strikers may induce violent acts of assault, threats, and sabotage; a strike of any dimension has the potential to create havoc and turmoil.

Civil disturbances, unruly crowds: May lead to or become a cause of a riot, enticing people to loot, destroy business and residential property, and cause arson; causes can range from political motivation to a perceived unfair treatment of a person or group.

Public utility or public services disruption: Electric brownout or complete electric.

Communication/satellite failure: Landline and wireless; to electronic systems as well as telephonic and computer communications.

Chapter 16

Accidental Occurrences and Medical Emergencies

Although accidents and injuries will occur to any institution, the loss-prevention manager should be aware and prepared to render the necessary assistance during any incident that may take place on board. The more serious the incident, the more liability may be incurred against the business establishment, and it is incumbent upon the management of a company to be ready and able to investigate and defend itself against any civil or criminal litigation. This may include lawsuits concerning premises liability, product liability, and general liability caused by the action (or inaction) of an employee.

Remember that when an establishment is open for business and gives license and privilege to persons to enter therein, those persons have an expectation that the premises they enter upon will be safe and free of any egregious encumbrances or any preventable incident that will place them in danger.

Moreover, it would do well for any business establishment or institution that caters to and welcomes visitors, customers, or clients to have some employees, most preferably the company's security officers, at least trained in basic first-aid procedures. For an employee to stand around a stricken subject doing nothing might appear disturbing to onlookers, although there is no requirement that anyone, trained in first-aid or not, must attend to or aid an injured person. For further information on this topic, see Chapter 17 regarding the ethics and certifications controlling aid to an injured person.

However, under the "good Samaritan rule," without aiding and attending to an ill or injured person first-aid training owes the injured party a duty of being *reasonably careful* (calling for an ambulance, attempting to stop the flow of blood loss, comforting the injured party). If the injury or illness is made worse in some circumstances because of gross negligence or poor handling, thereby causing more of an injury, the injured party may pursue civil litigation.

For further information, see Chapter 28, "Familiarization with Criminal and Civil Litigation."

Chapter 17

Infectious and Health Hazards

Other than the terrorist or criminal act of dispersing bacterial or viral agents covered elsewhere in this book, the accidental transmission of bloodborne pathogens must be considered a major health hazard. In addition, the unintentional workplace incident where some hazardous or toxic substance might affect anyone in and around the business facility must be considered.

Bloodborne Pathogens

Because of the possibility of contamination from bloodborne pathogens, the first-aid responder must be aware of the risk that he or she may face while rendering first aid to an injured person. Moreover, there is a risk of transmitting the infection to the responder's family or to others. Pathogens include blood and body fluids that carry HIV leading to what is commonly known as AIDS. Hepatitis A, B, C, and other viral infections can also be transmitted via bloodborne pathogens.

The federal Bloodborne Pathogen Standard was designed to protect workers from the risk of exposure to pathogens that can cause chronic diseases such as those noted above. The standard (law) requires the following:[38]

Exposure assessment: The employer must determine which job classifications and specific tasks involve exposure.

Written exposure control plan: A site-specific exposure control plan must be developed and be in effect.

Hepatitis B vaccination: Employees who are at risk to this exposure must be offered free hepatitis B virus (HBV) vaccinations.

Education and training: The required bloodborne-pathogen training sessions for employees at risk must be offered by the employer.

Recordkeeping: Medical records of employee exposures and treatment must be retained and kept confidential for the duration of employment plus 30 years.

Hazardous waste disposal: Safe cleanup and disposal methods must be used, with appropriate biohazard labeling.

Post-exposure incident management: The employer is responsible for initiating the required actions after exposure of an employee, managing after-care for the employee, and maintaining continuity of records.

With appropriate first-aid and cardiopulmonary resuscitation (CPR) training, the security officer and other applicable employees will know the correct procedure of protecting themselves and others. All human blood and body fluids must be treated as if known to be infectious. As long as a barrier is placed between the victim's blood or other body fluids and the first aid responder, and there is the careful disposition of blood, bandages, and other emergency items (proper infection control methods), there is little chance of contamination to the provider.

However, an employee may fail to give aid because of a repugnant sight or because of fear of certain (infectious) conditions, and may not want to expose himself or herself (and ultimately his or her family) to any transmittable disease under any circumstance. Other than health care professionals such as physicians, nurses, certified EMTs, etc., who are controlled by ethics or certification, there is no law known to the author that requires a first-aid responder, trained or not, to aid an injured person. That decision is left to the individual.

Nevertheless, a civil action could possibly be initiated if an officer fails to act in the emergency that he or she was trained for and, in substance, would be an action *against the employer rather than the officer.*

> There is no legal requirement for a person, whether trained in first-aid procedures or not, to render aid, and no civil recourse by a plaintiff can be justified.

The failure to act in an emergency situation has no bearing on the various state statutes regarding the "good Samaritan rule" or laws about aiding an injured person, in which anyone acting in a reasonable way and without gross negligence will not be held responsible for any harm that may occur.

Security officers, particularly those certified as EMTs, etc., should check their own state statutes to ascertain any requirement or procedure that must be met in aiding an injured or ill person.

See Appendix B for information in developing an acceptable bloodborne pathogen exposure plan.

Hazards in the Workplace

There is another OSHA (Occupational Safety and Health Administration) law that must concern all security and safety officers because of a possible health hazard: the Hazard Communication Standard "Right to Know Law." All employees have the right to know what hazards or harmful substances they might face or may come into contact with in their workplace. This includes all chemicals, solvents, cleaning agents, toxic substances, and environmental hazards. This law also includes the inventory, labeling, training, and documentation of all employees.[39]

Not only security and safety officers, but also all employees should be aware of all the hazards that may be contained in the workplace:

A ***hazardous substance*** or mixture is any chemical that is a physical or health hazard.

A ***physical hazard*** is any item that is explosive or flammable, including combustible liquids and compressed gases.

A ***health hazard*** is any chemical substance that—when absorbed through the skin, inhalation, or ingestion—may produce acute or chronic effects.

For a full description of this law, see "Hazard Communication Standard" in Appendix B.

Hazardous Materials

Potential Hazards

Substances or materials in any form or quantity that pose an unreasonable risk to safety, health, or property include those that are:

- Flammable
- Toxic
- Corrosive
- Reactive

International Classification System of Hazardous Materials

The following classification is accepted internationally, particularly regarding the transportation and storage of these dangerous materials:

Class 1: Explosives (TNT, nitroglycerin, etc.)
Class 2: Gases (flammable, nonflammable, poisonous, corrosive)
Class 3: Flammable liquids:

Division 3.1: Liquid with a flashpoint below 0°F
Division 3.2: Liquid with a flashpoint above 0°F but below 73°F
Division 3.3: Liquid with a flashpoint above 73°F
Class 4: Flammable solids (spontaneous combustible materials, materials dangerous when wet, metals)
Class 5: Oxidizers and organic peroxides (common pool chemicals)
Class 6: Poisonous and infectious materials (bloodborne pathogens)
Class 7: Radioactive materials
Class 8: Corrosive materials (battery acid, sulfuric acid)
Class 9: Miscellaneous materials

Hazardous-Materials Labels

Basically, any hazardous or toxic substance within a facility must have a label to advise and warn employees who may handle or come into contact with such material. The label will note the following in particular:

- Name and address of the manufacturer
- Identity of the material
- All appropriate warnings

All materials that fall under this standard and enter a facility must include a material safety data sheet (MSDS) provided by the manufacturer or distributor. These MSDS documents must be on file and available to any employee for inspection and review. Each MSDS must contain the following:

1. Specific chemical name
2. Common name
3. Physical and chemical characteristics
4. Health effects following exposure
5. Related health information
6. Exposure limits
7. Carcinogen warnings, if applicable
8. Precautionary measures
9. Emergency and first-aid procedures if exposed, following spills, etc.

Chemical and Toxic Spills

This category includes hazardous materials and any substance that is harmful or toxic and poses risk to people and the environment, even when handled, used, or made in a reasonable way. Such materials may be of a chemical, toxic, or flammable nature that can cause serious health or environmental hazards. Occurrences are generally accidental exposures from spills, leaks, or unintentional contact. Personal contact can occur by inhalation, ingestion, or skin exposure (topically).

Emergency Procedures

If a spill is of some consequence, contain and secure the area around the spill as soon as practical. Additionally, evacuate all personnel and occupants from the contaminated zone to prevent exposure or injury from explosion. Large spills must be contained and cleaned up by public authorities that have the equipment and expertise to handle such a situation. This most often will include police or fire hazard material (hazmat) units. The extent of a small spill or leak within the premises of a business and how the security officer or personnel on board handle it depend on the seriousness of the incident and the safety procedure to be followed.

For small spills, where an employee or employees have been completely trained in hazardous-materials conditions, possess the expertise, have the correct protective equipment and the proper hardware to control, contain, and clean up the hazard, they may do so with care. If a hazard exists that will cause harm or be a danger to employees, the public, or the environment, public authorities must be contacted.

Prevention

- Good housekeeping is essential.
- The security and safety officer must be constantly observant for possible problems or accidents.
- These officers must be completely familiar with the Emergency Procedure Plan, which should describe proper steps to be followed in any hazardous-material accident.

Chapter 18

Fire Emergency[40]

Fires cause more death, injuries, and property destruction than any other threat. This is most illustrated in the residential area, where children and the elderly fall victim to death and injury more frequently than in a business situation. Fire safety and firefighting procedures are more practiced in the business community than in the home, and therefore, once a fire is discovered, correct response reduces injury and damage.

Whether at home or at work, statistics have shown that fire is the number one cause of death and property destruction, and fire professionals are convinced that most fires are caused by poor housekeeping procedures. Moreover, many fires may be caused by oversight and mechanical, engineering, or appliance failure. If the fire incident is classified as a negligent act or an act of omission, criminal or civil action may result against the person, persons, or company involved. We must also consider arson as a criminal act with intent to cause damage.

Nevertheless, whether the cause of the fire is natural, accidental, or criminal, the results can be devastating to a business enterprise. Each business may have particular fire risks that must be addressed. Retail department stores contain warehouses, storerooms, and closets that house far more combustibles and are more exposed to fire than supermarket or grocery stores. Plastics, wood, wood composition, and plastic foam contained in furniture, bedding, and pillows ignite and melt easily, generate high heat, and produce toxic fumes and heavy smoke. High-piled palletized stock may be very flammable. Institutions such as hospitals contain highly explosive and flammable liquids and gases, and buildings such as libraries may be considered high risk because their contents can burn quickly and easily.

Sprinklers in warehouse racking are considered a necessity, and fire codes following national guidelines set by the National Fire Protection Association (NFPA) are generally mandated in all public buildings. Fire-resistant buildings, completely sprinklered throughout the facility and constructed in sections with fire containment walls and fire doors between major portions of the building, can contain or deter the spread of fire until the arrival of the fire department. The local fire department should have the availability to conduct routine walk-through inspections, particularly during business hours. This is most important during routine construction and renovation. Local fire or building codes will dictate the amount, type, and placement of fire extinguishers and other relevant hardware. Such codes will also note the number and type of fire exits that will be required, how and when they are or are not to be secured, and how all perimeter doors will operate.

Good housekeeping is one of the most important fire prevention methods. Clean and orderly areas throughout the building can only reduce the probability of fire. But fires do happen. From a tilted lampshade that comes in contact with a hot electric light bulb to an electrical short in a fixture, building personnel must be prepared to react swiftly.

Another area of great concern is the danger of fires that involve or infringe upon electrical transformers. There is a high probability that, in such an occurrence, the transformer contains PCBs (polychlorinated biphenyls). As toxic smoke from a fire of this type invades the facility's air conditioning or heating system, everything it touches will be contaminated and must be sealed off until a contractual decontamination team can clean it up and advise when the building is safe to reoccupy. Realize that a cleanup may take weeks or months, depending on the extent of the contamination.

A similar danger can occur in many chemical spills where chemicals could mix in a fire or even create fires themselves. Fires of this kind, where a chemical is contained in and is part of the fire, will require special training and self- contained breathing apparatus for members of the fire brigade. Without this equipment, security officers should evacuate the area and contain the fire as much as possible without causing harm to themselves.

The Effect of Fire upon People

Individuals respond differently in emergency situations. No matter how much training, fire protection, and fire equipment may be brought to or contained in a facility, we cannot control people once a fire has started. During any fire, human behavior can be unpredictable, with the primary objective being able to escape safely, hopefully with their family members. In many instances their actions may be considered erratic under the circumstances, such as attempting to flee the scene with their possessions in hand. Studied behavior patterns by gender and age may be considered a good indicator of a reaction to fire and smoke, but behavior can change from fire to fire, with the primary objective of survival for the victim. Whatever the facility, the consequences of a fire are most serious because people can resist fire and its by-products only for a very short period of time.

Panic

Then there is the factor of panic. If they are unable to see or exit freely and safely, people may become frantic in an effort to save themselves. Many die by piling up against exit doors and windows or are trampled by the rushing mass of people in their frenzied attempt to escape. Generally speaking, because of panic, superheated air, or smoke, people can die even before the fire reaches their bodies.

Depending on the type of fuel involved in the fire and how fast the fire is spreading, the first indication of effects will be lack of oxygen combined with smoke, toxic gases, and chemical compounds. Fire uses up oxygen faster than the victim, and suffocation leads to death. Dense smoke becomes an irritant to the eyes, lungs, and air passages. It causes the eyes to water, the nose to run, and the throat and lungs to cause coughing and vomiting. The most common toxic gas is carbon monoxide, which, when absorbed into the body, will cause brain dysfunction. The person becomes confused and disoriented and can succumb to carbon monoxide poisoning without ever seeing the fire. The final factors are heat and superheated air that blister and destroy tissue. When air temperatures reach 200°F, the chances for human survival are minimal to none. A person cannot conceive the effects on a burn victim who has survived a fire in the pain, suffering, and permanent disfigurement that can occur.

The toll in human tragedy in deaths and injuries, property damage, and the disastrous aftereffects that follow a major fire can have a profound effect on both those involved and the community at large. Therefore, the importance of fire protection cannot be overstated. Protection of the facility with fire-resistant materials, fire hardware, training, inspection, and enforcement is the key to life safety initially and, following that, property protection.

> Remember that fire prevention is proactive, and fire suppression (the response) is reactive.

The Fire Command Station

The fire control panel or fire command station is considered a *proprietary system* and must be manned during the hours of the business operation. This station may be located at the company's switchboard, security office, or any location where the monitoring person has knowledge of the operation of the system and the appropriate response to an alarm. Whether the system provides a digital display, an audible local alarm, or a printout of the transmitted signal, it may also be programmed to shut down heating, ventilation, and air-conditioning units (HVAC), turning on exhaust fans, initiating the closing of fire doors, and directing elevators to bypass floors and return to the lobby. The person who mans this station will abide by the fire safety plan, make the proper notifications, and respond to members of the fire brigade for whatever assistance they may require at the scene.

The off-site central station system receives the alarm signal from the fire control panel, but the employees there have no direct knowledge of what may be happening within the facility other than they must assume the incident is a fire. If this system goes into effect after the close of the business day, any signal from a smoke or heat detector, sprinkler-head water-flow detector, or a tamper alarm, for instance, would require notification to the fire department for its immediate response. Most central station systems monitor on a 24-hour basis and, upon alarm activation during business hours, will first call the subscriber to determine if there is a fire and whether the fire department is needed. The reason for this form of notification is that, depending on the type of business and the number of occupants other than employees, false alarms can be quite frequent.

The fire command station may be included in or made part of the disaster command station noted elsewhere in this book, in order to simplify a comprehensive control location.

Fire Strategy and Training

Every business establishment should have a fire safety strategy as part of the Emergency Procedure Plan in effect for all employees to follow. This plan should include what an employee should do upon first observing a fire and how to report it, and how and what notifications are made to all employees so as to make them aware of a possible emergency and possible evacuation. Additionally, the plan should note the procedure of reporting the fire to the fire department, and the response of an internal fire brigade or other group of trained employees to respond to the scene so that they may extinguish or contain the fire until the arrival of the fire department. If the fire brigade determines that the fire is of a serious nature or is out of control, an evacuation procedure for all inhabitants of the building should be placed in effect.

Whether your company does or does not employ the services of a fire safety director, and if the facility in question is a high-rise office building, hospital, hotel, or school, the assignment of a fire warden to each floor with responsibilities defined in the fire safety plan would be considered an important part of fire safety. Moreover, some municipalities require that the fire operations plan be approved by the fire department, and that certain businesses must have certified fireguards on duty at all times of the day. These fireguards, who are also part of the fire brigade, will have special training in fire protection and firefighting over that required of security officers or other members of the fire brigade. Fire inspection duties would be a major part of the fireguards' responsibilities. The most important and primary obligation is to observe and report all fire conditions and hazards.

The security officer should also be an integral part of the fire safety team. Many businesses or municipalities require that security officers also be certified as fireguards. In large business establishments and institutions, the security officer has the additional duty of fire inspection as part of his or her inspection and auditing functions. This includes the routine inspection of all fire hardware such as pull stations, smoke and heat monitors, sprinkler systems, fire extinguishers, fire hoses, and fire exits and passageways. Also included is the inspection of the proper storage of paints, cleaning agents and solvents, gasoline and kerosene, along with monitoring good housekeeping procedures.

In any large company there should be a written fire safety plan, which will describe the duties of certain building personnel and how all employees are to respond to a report of a fire. This plan would be particularly beneficial if company employees are located on more than one floor or level or in a high-rise building.

In particular, the security officer should be completely aware of the company's emergency procedure in responding to a report of a fire. This will include what everyone's duties are concerning all aspects of a fire safety plan, which are noted above.

All employees, and especially members of the fire brigade and all loss-prevention/security and safety officers, should be trained in the following:

1. The science of fire; the chain reaction of heat, fuel, and oxygen
2. The classifications of fire
3. How to prevent a fire
4. What fire extinguisher or agent to use on each type of fire
5. How to fight a fire
6. The consequences of a fire on a business, people, and property

Chapter 19

Fire Science and Fire Classification

Most security professionals are trained in security matters rather than safety. Consequently, they may not be as familiar with fire—a major safety threat to the business environment. The information in this chapter is presented in an effort to broaden the security officer's conception of security to include safety when faced with the serious hazard of fire.

Fire Science

The Four Stages of Fire Development

1. *Incipient stage* (minimum hazard): Invisible particles generated by thermal decomposition; no smoke, flame or appreciable heat
2. *Smoldering stage* (moderate to major hazard): Combustion products now visible as smoke; flame or appreciable heat still not present
3. *Flame stage* (major hazard): Actual fire now exists; appreciable heat still not present, but follows almost instantly
4. *High-heat stage* (extreme hazard): Uncontrolled heat and rapidly expanding fire; severe damage

The above stages describe how a fire will evolve from a minor flame to a conflagration that could destroy a business as long as the chain reaction elements of heat, air (oxygen), and fuel (burning material) are present.

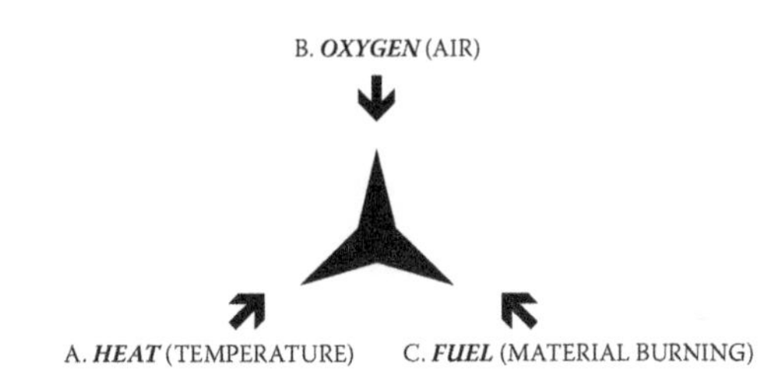

Figure 19.1 Chain reaction of heat, oxygen, and fuel.

Fire is the result of a *chain reaction* of heat, oxygen, and fuel (see Figure 19.1). A fire will continue to burn (the chain-reaction sequence) until one of these three elements is removed:

Heat: The heat is removed by cooling with water or CO^2 (lower the temperature).
Oxygen: The oxygen is reduced or smothered by foam, CO^2, dry chemical agent, or covered in some manner (smother or remove the air).
Fuel: The fuel (wood, paper, oil, gasoline, etc.) is either totally consumed or removed (starve the fire).

The Classifications of Fire

The security officer should know the four classifications of fires and the types of extinguishers that should be used for each type. The use of the wrong fire extinguisher may be of little use and could cause more damage.

1. ***Class A***: Ordinary combustibles such as wood, paper, cloth, and rubbish. Normally extinguished by cooling down the burning fuel; use water, soda acid water tank, or CO_2 to extinguish.
2. ***Class B***: Flammable liquids such as gasoline, oils, solvents, paints, and cleaning agents. Best extinguished by smothering; use CO_2, dry chemicals, or foam. Do not use water, since it will cause the fire to spread rather than diminish.
3. ***Class C***: Electrical or electronic fires such as burning wires, electric motors, or circuits; materials with live electrical current flowing. Do not use water or foam; to prevent electric shock, use a nonconducting extinguishing agent, such as CO_2 or dry chemicals.
4. ***Class D***: Combustible metals such as magnesium, sodium, lithium, iron filings, aluminum chips, etc. The extinguishing agent is dry chemical powder. Class D is a very hot and precarious fire, and usually occurs in manufacturing facilities that deal with these types of metals.

Remember that water cools down a fire, CO^2 cools and removes oxygen, and dry chemicals cool and removes oxygen.

Firefighting Equipment

Extinguishers come in varying sizes and shapes:

Water hose: Typically located in areas that are susceptible to a Class A fire (receiving/shipping docks, warehouse areas, offices)
Metal canister: May contain water or soda acid under pressure for Class A fires or CO^2 and dry chemical for Class B and C fires

The most common hand-held fire extinguisher used today in most facilities, public and private, is a metal canister Class ABC unit (dry chemical, multipurpose), which is applicable in fighting three classes of fires. Unless a person about to fight a fire has the availability of an ABC fire extinguisher, which can be used on all four classes of fire, the correct fire extinguisher must be used.

Conclusion

The loss-prevention manager or safety manager must be fully informed of the types and use of assorted fire hardware contained within the facility. This will include the types of hand-held fire extinguishers, fire hoses, sprinkler systems, standpipes, fire alarm pull stations, smoke and heat alarms, and the fire activation system. Although the maintenance and required inspections of these systems and hardware usually fall to the fire safety director, engineering, or the maintenance department, such duties may be a part of the loss-prevention manager's range of responsibility.

As noted, fire experts have concluded that poor housekeeping practices within a facility cause most fires. The importance of routine documented inspections and consistent daily patrol by security officers cannot be overstated when it comes to fire safety. Any observed violation or poor practice must be attended to immediately.

Although most fires within a business enterprise are accidental or result from poor housekeeping practices, we cannot overlook two other types of fires that can cause great harm and destruction:

Wildfires: Natural, accidental, or intentional fires of a serious nature in woodlands encroaching upon residential and business areas
Arson fire: An overt criminal act for the purpose of destruction, and which may cause death and serious injury

Chapter 20

Fire Safety Procedures and Guidelines

The Fire Safety Director

The fire safety director should have complete authority to organize and assign building personnel to the fire brigade. He should also be responsible for a written fire safety plan, the state of firefighting capabilities, fire and evacuation drills, and the designation and training of fire wardens for each floor or large areas within the building. This position will have complete authority over all occupants (building service, tenant personnel) and the various company entities that may inhabit the building.

His or her other duties and responsibilities will include the routine inspection of all fire fighting equipment and hardware, and the complete supervision of the fire command station for the coordination of firefighting strategies and evacuation. In his or her absence, a responsible and trained subordinate will assume these duties.

If the building, institution, or company in question does not employ a fire safety director, all duties and responsibilities shall devolve upon the loss-prevention/security manager.

Assignment of Safety Personnel

Depending on the size and type of the building, the business property, and the number of employees and occupants therein, the building and/or company personnel should be assigned the following roles:

Fire Warden

The fire safety director may also hold the position of the building fire warden. The fire warden should:

- Be thoroughly familiar with the fire safety plan.
- Be thoroughly familiar with all fire exits, fire extinguishers, and fire hoses.
- Know the number of occupants in the area of responsibility and formulate a traffic pattern for evacuation through emergency exits.
- Conduct daily visual inspections of all fire equipment and hardware. Also inspect all fire exit signs and emergency lighting to verify that they are in good working order.
- Conduct daily visual inspections of all fire doors and fire stairways to verify that all fire doors are kept closed, are not locked, and that doors and stairways are free of any obstruction. A subordinate may be assigned to conduct these visual inspections.
- Ensure that all occupants under the warden's control are notified of a fire report. In the event of a fire report, the warden should immediately initiate the evacuation of all occupants from the floor or area under his or her control in accordance with directions from the fire command station.
- Be aware of all employees who are handicapped or disabled in some manner, and who may need assistance during an evacuation. The warden should maintain an up-to-date list of all personnel who fit into this category. A copy of this list should also be maintained at the fire command station.
- Direct all personnel to the safest stairway based on information from firefighters, the fire command station, or personal observations.
- Before leaving, determine that all spaces and areas have been examined and cleared by the searchers.

Assistants to the Fire Warden

The following employees will be assigned to assist the fire warden:

Deputy fire warden: An assigned deputy or assistant fire warden will assist the fire warden during the emergency and will assume the responsibilities of the fire warden in his or her absence.

Floor wardens: Depending on the proportions and number of floors contained in the facility, a building employee or a company employee may be assigned as a floor warden. Each floor of the building should include the assignment of a floor warden or, in other circumstances, a deputy floor warden may be assigned to a floor occupied by a particular business and under the direction of the building fire warden. However the designations of floor wardens are made, the number of floor wardens assigned should depend on the square footage of occupied space on each floor or area.

Searchers: Designated employees should be assigned as searchers to each floor or area. Their duties are to check all offices, conference rooms, rest rooms, break or lunch rooms, storage rooms, reception and lobby areas, and any other spaces or remote areas in order to notify, advise, and ensure that all occupants are evacuated.

Monitors: For those personnel who have disabilities, monitors shall be assigned to assist and help in the evacuation from the premises. During a fire emergency, they are not to use an elevator for evacuation unless assisted and manned by a professional fire fighter.

Stairway monitors: During a fire emergency evacuation, stairway monitors shall take their position at their assigned stairway. These persons shall immediately:

- Inspect the stairwell for smoke and heat before and after the evacuation of personnel
- Instruct personnel to form a single line down and to the right side of the stairs
- Attempt to control any panic or hysteria
- Determine that all searchers have cleared all personnel and visitors
- Make certain that no one uses an elevator for evacuation unless directed by a professional fire fighter

During any emergency or evacuation procedure, managers, supervisors, and those employees assigned to a specific duty, as noted above, must act in a responsible and professional manner. To act otherwise—exhibiting nervous, agitated, or irresponsible behavior—could cause fear, apprehension, panic, or hysteria among other coworkers and building occupants.

See Appendix A for a sample fire safety plan that may be implemented by an institution or a business establishment.

SECTION 4

Chapter 21

The Disaster Management Process

Introduction

A business exists to make money, and an institution serves a specific purpose, whether public or private. When a disaster affects any enterprise, it can cause a major loss of productivity and service that will affect the bottom line. The ability to identify and respond to a disaster or emergency is one of the more important roles for the loss-prevention department and the security and/or safety officer. Training, experience, and company policy as well as local, state, and federal laws should be taken into consideration where appropriate and relevant.

Business Disruption

A disaster or serious emergency can occur in many forms. It can be sudden and without warning, such as a terrorist act or an earthquake, or it can allow some forewarning, such as observable employee behavior prior to a serious assault upon others. Generally, though, for other than natural occurrences, we tend to become complacent and are often indifferent to indicators that our business could suffer some degree of destruction, or the employees and the public we serve may become victim to death or injury.

In addition, a business disruption can occur in several ways and in varied degrees. Other than the actual event or a threat of bomb activation, it may include other human-made disasters such as arson, sabotage, and espionage as well as natural occurrences that cannot be discounted, since the subsequent damage or injury to people and property can have an everlasting effect on the business entity.

A disaster is an unplanned occurrence that disrupts the routine operation of a company. It may include one or more of the following occurrences:

Terrorism: An intentional act resulting in death and destruction, and in effect, causing fear, panic, and apprehension, usually with the use of a bomb device; such an act is committed for various reasons, which may include political, religious, cultural, or ethnic motivation.

Sabotage: The intentional damage to property in an effort to halt or slow down a business operation or a government entity.

Espionage: the theft of proprietary information, business plans, or confidential records that will benefit a competitor or another and may do harm to a business entity or a government entity.

Power outages: Brownouts or blackouts that affect the business operation and the mobility of the populace.

Computer damage, computer theft, system crashes: Any intentional damage or theft of operating systems, programs, business information, or system crashes that can cause serious interruption to the business operation with long-lasting effects.

Natural or accidental occurrences: Natural disasters such as fire; excessive rain, snow, or flooding; and major disasters such as tornadoes, hurricanes, and earthquakes, all of which can cause death, serious injury, and severe property damage and disrupt the business operation. Unless the disaster is known beforehand, there is little time to analyze how severe or extensive the effect will be on the business in question. Earthquakes and tornadoes occur without little warning for anyone to prepare any protection for self or property. Hurricanes, severe rain, or flooding will offer a period for analysis in preparing for the occurrence. Depending on weather reports and other notifications, the loss-prevention professional usually has some time to employ resources in an effort to protect people and property as much as possible.

Violence: An assault in the workplace that may or may not include a weapon; assaults may be broken down into two categories:

Internal: An assault causing death or injury to employees and managers by another employee or former employee. An assault of this type may be predictable to some degree if there are indicators of employee behavior that are observable and, once identified, some action is taken. Be aware that an assault on or around the premises may also include domestic violence and sexual attacks, in that the participants may know each other and that the occurrence may result from events outside the premises.

External: An assault causing death or injury to the inhabitants or employees of a facility by a disgruntled or mentally disturbed person, who may have been a former customer, client, or patient. An incident of this type is ordinarily unpredictable, since the action is usually a spur-of-the-moment act or an act of mental instability. This type of assault may be directed toward a particular person or group, or it may occur as a random act within the business facility, as the company may be the target of anger. Robbery may also be included in this category. A robbery may be classified as an act by one or more perpetrators, on or off premises, with or without a weapon, during the theft of property from someone by force or threat, and where death or injury has or could have occurred.

The Execution of Violent Acts

Every country is vulnerable to international and national terrorism. Along with that, any business establishment or institution may be subject to acts of violence by a disgruntled or deranged individual. Vistors and employees may be killed or seriously injured with the possiblility of severe property damage. Anyone may be the subject of a violent or terrorist act, whether the perpetrator is foreign or domestic.

Consider also, to a great degree, that incidents of this type cannot be anticipated. But whether the act is planned or random in nature, business executives and loss-prevention managers can employ measures to prepare for and respond to such events, thereby helping to prevent death, injury, and damage in some measure.

The American public must suppress the mindset that it can not happen here. The shootings at CIA Headquarters in Virginia in 1993, the Oklahoma City bombing of the Murrah Federal Building in 1995, the Centennial Olympic Park bombing in Atlanta, Georgia, in 1996, the Pentagon in 2001, the World Trade Center in 1993 and its complete destruction in 2001, and the various bombings at abortion centers give indication that these tragic incidents, which have caused so much death and destruction, will most probably happen again.

Law enforcement and other public safety groups have the responsibility, resources, and ability to effectively mobilize and respond to a disaster or emergency, no matter what type or how serious. Generally, the private sector is much less trained and organized to respond to any type of serious emergency or disaster.

Prevention may be the most difficult objective when it comes to a terrorist or violent act. In reality, we are dealing with an unknown, and many times an unpredictable escapade. Because of the nature of the business establishment where access by the public is the norm, it becomes much more difficult to prevent these unlawful acts from occurring in the workplace. So then, how do we attempt to prevent criminal acts against our facility and prepare for those occurrences that may be unavoidable?

The answer may be as simple as ***knowledge***, ***preparedness***, and ***training***.

Command Structure

Command Manager or Director

The term *command structure*, as described in the disaster plan, should be definitive in nature to include a *command post* or *command center*, the purpose of which is to centrally control responders; link individuals, groups, and agencies; and assign personnel, emergency equipment, and available resources as needed, with all decisions and authority at their command. The ability to assess the seriousness of an incident, determine the central objectives, and prioritize the type and level of the response, and then to supervise and act as necessary, are extremely important skills for the coordinator. Therefore, proper assignment of personnel to this post, and the training of those personnel, is critical.

In particular, a responsible person with the training, ability, and knowledge of the facility and the resources available, who can take control of any situation, no matter how hectic or immense it may be, should be assigned as the command manager or command director at the command center. A hierarchy of command should be written into the company's policy and procedures so that, whatever emergency may arise, the personnel assigned to the command center is delineated.

For the purposes of this book, we will assign the title of the person who will be in charge of the command center as the *command manager*. During any emergency procedure, the person detailed as the command manager must have the *ability* and *authority* to ***organize***, ***deputize***, and ***supervise***.

Remember that in very serious occurrences, and depending on the type of emergency, public safety officials will take command upon their arrival, with company personnel assisting them as required.

Concerns

The most important question that the management of loss prevention personnel must ask is, "Can this happen here?" Consideration should be given to any and all emergencies, disasters, or violent occurrences that could occur on, within, or near the business establishment.

In regard to the possibility of a catastrophic occurrence, the intent of loss prevention is to examine, identify, prepare, plan, test, and manage.

Operational Objectives

As noted earlier, if the business enterprise is of some stature, a written disaster plan should be created so that the company may specifically address the possibility of an emergency that could cause disastrous circumstances. Any plan that details the response to various disaster or emergency incidents should clearly state that such response is made for the protection of life and property. Following that is the resumption of the business operation with as little disruption and loss as possible.

Initially, we must:

1. Understand what constitutes a disaster.
2. Know how to respond to disaster situations.
3. Know who is to be involved in the disaster response.
4. Know what duties and responsibilities have been assigned to these responders.
5. Train and drill for disaster response.
6. Have a plan for implementing a stabilization and recovery process.

Disaster Preparedness Considerations[41]

In order to be prepared, we must ask ourselves: What would we do if a disaster or major emergency were to occur? How would it affect my company, my personnel, or anyone else at the time of or after the occurrence? How severe will the incident be, and how much of an effect will it cause even in the partial or complete disruption of my business? And finally: What if the worst did happen? Could the business survive days, weeks, or even months of closure?

Disasters and emergencies do happen, and how severe and how prepared we are to address these incidents will determine the viability of our business enterprise.

What If?

Operations

1. What if there was a prolonged electric power outage or a disruption in natural gas service?
   - Immediately notify the appropriate public utility for repair or alternative services.
   - Because blackouts and brownouts are becoming more common, alternative sources must be considered. A business, should have a backup generator to maintain full power, or at least adequate power to provide for the critical functions of emergency lighting, refrigeration, security systems, and computer control.
2. What if there was a bomb explosion or serious fire?
3. What if there were substantial and serious injuries to occupants?
   - Initiate procedures as detailed in the Emergency Procedure Plan.

- Immediately notify public safety authorities for their assistance and provide complete cooperation upon their arrival.
- Be guided by fire, police, and building safety inspectors for authority to reoccupy full or partial areas of the building.
- Determine the structural and physical safety of the building before reoccupying and then resuming partial or full operation.

4. What if the facility were closed for a period of time or completely destroyed?
   - Develop contingency plans to remain in operation off-site if possible. Can some equipment, inventory, supplies, and computers be salvaged and reactivated on- or off-site?
   - Initiate recovery and restoration procedures as soon as practical.
5. What if my suppliers and shippers suffer a shutdown even though my company is still in operation?
   - Maintain hard-to-replace parts and supplies on hand and stored off-site. If not practical, achieve cooperation with suppliers and shipping companies in advance in ensuring a secure and adequate supply when required. Relationships and agreements should be made in this regard prior to any incident that may occur.

Information and Communications Systems

6. What if my computers and peripheral systems, or my computer programs and critical software, were destroyed?
   - The company should provide for a computer incident response team (CIRT), a group of professional computer programmers and technicians, in-house or not, that can respond to an immediate computer emergency. It would be well advised that surge protection for all computer and telephone systems be installed so that a power surge cannot interrupt or destroy these connections.
7. What if my computer and/or business records, including on-site backups, were destroyed?
8. What if payroll, tax, accounting, stock/inventory control, or production records have been destroyed?
   - Backup copies of all computer programs, computer records, and all business records must be maintained and upgraded on a routine basis. On-site copies should be contained in secured fire-safe cabinets. If required, a second set of backup copies of the more important records should be maintained off the premises.
9. What if there was a telephone or complete telecommunications termination?
   - As to telephone problems, immediately notify that utility requesting restoration and repair or alternative service.
   - Once telecommunications are restored, make sure as part of the Emergency Procedure Plan that an up-to-date list is compiled of all important

telephone numbers of public safety authorities, utilities, security services, contractual services and administrative and key employees to be notified depending on the type and severity of the occurrence.

Insurance Coverage

10. Is my company's insurance coverage adequate to cover all risks, including civil liability? Am I aware of the deductibles?
    - Current insurance coverage must be reviewed and upgraded on a routine basis. Plan for the worst; a disaster or emergency may never have taken place prior, but may occur without warning. Be prepared with the necessary coverage. Be thoroughly aware of the insurance coverage that protects your company.
11. Is our insurance adequate to cover our damage, present and expected losses, replacement, and business interruption? Is it broad enough to put the business back into operation after a disaster?
    - Insurance coverage may be broken down into the following classifications:

      General liability (operations and premises): protection against claims made by people not employed by your business who are involved in an accident or an incident. Depending on the incident, it may include an employee.

      Property: coverage for fire damage and other natural or accidental incidents. Be apprised that regarding coverage for property damage, death and injury caused by a terrorist act or a wartime occurrence, an underwriter may elect to reduce or refuse insurance coverage, and will be so noted in the insurance policy. See also "Risk Versus Uncertainty" in Chapter 22.

      Auto: coverage for company owned or operated vehicles that become involved in accidents.

      Workers' compensation: coverage for employees who are injured on the job or in the course of their employment.
12. Do I, as the risk- or loss-prevention manager, understand what is covered and what is not? Are extra riders needed to cover certain risks?
    - Maintain close contact with your company and the insurance carrier's attorneys, the insurance carrier or insurance broker, and your carrier's insurance investigators.
13. Do I have access to insurance loss data and do I understand the contents?
    - Maintain close contact with two of the more important insurance representatives because of the services they can provide:

      Your insurance company's *loss-control representative*:
        - Helps develop and implement an effective safety program in an effort to reduce accidental loss
        - Reviews and advises in the strengths and weaknesses affecting all your assets and the health and safety of the public and your employees
        - Can prescribe accident and loss control procedures and systems

- Offers advice regarding safety laws, codes, regulations, and standards to your business operation
- May attend and give input at safety committee meetings

Your insurance company's *claims representative*:

- Investigates claims to determine compensative coverage
- Maintains a cost-effective claim service
- Evaluates medical information, attends medical settlements, and reviews medical costs of health care providers
- May attend hearings, settlements, and attorney conferences on your behalf
- May settle claims and set monetary reserves
- Will initiate legal representation for your company for any civil litigation as covered under your policy

In Summary

A good safety and security program begins with a substantial and independent assessment of the following factors:

1. ***Inherent risk*** to the
   - Facility
   - Employees
   - Visitors, clients, students, patients, and customers
   - Institution or the business enterprise itself
2. ***Vulnerability*** of each of those elements noted above.
3. ***Necessary preparation*** regarding reaction to the incident:
   - The required safety and security equipment, hardware, and devices needed to protect the facility and its contents (people and property)
   - Assignment of responsibility
   - Training of the responders (security, fire, and safety)
   - Safety training for all employees
   - Routine drill and practice, and the documentation of such activity
4. ***Review and inspection*** of the physical plant
   - Determine that fire safety and building safety provisions and requirements are strictly maintained to code prior to any incident that may occur
   - Monitor and evaluate all programs, procedures, training, drills, etc.
   - Update and correct as necessary

Pre-Incident Planning and Assignment of Responsibility

The professional approach to effective disaster and emergency management is the rapid and well-planned response. The quality of a response to a disaster or emergency depends on the planning, training, and drilling that the business entity is committed to. This will include interaction and cooperation with all outside public safety and utility services.

Response plans must identify and detail authority levels and responsibilities, determine the resources present on site, assess the availability and timely response of resources off site, and identify public safety agencies for mutual aid, assistance, and support. This will include police, fire, ambulance, electric, gas, and telephone emergency response.

Planning for an incident must include all possibilities, with each level of seriousness correctly identified. Whether it is the anticipation of a bomb threat or explosion, or the serious assault by one or more people on the inhabitants of the facility, the inherent risk, sensitivity, and vulnerabilities must be prioritized, along with the present security measures and the need for correction or the enhancement of those procedures.[42]

Not all bomb incidents are the work of international or domestic terrorists; they may arise from a criminal extortionist or a disgruntled or mentally unbalanced person. Moreover, as we have seen in our recent history, emotional stress and mental instability can cause random death and injury by individuals in the workplace.

Planning[43]

In planning for a disaster response, the policies and procedures of that response must be made part of the Emergency Procedure Plan, as would a fire safety plan and any other emergency situation that should be addressed.

Along with those considerations noted under "Assignment and Support Services" in Chapter 22, the subsequent concerns must be part of the planning process and included in the Disaster Procedure Plan:

- Determine the key areas of protection and prioritize the areas of concern
- Determine if and when public safety services are warranted and are to be notified
- Identify key personnel who will respond to various emergencies or occurrences, and their responsibilities
- Assign the person who will take charge of the command post
- Designate the location of the command center and the responsibilities of any personnel assigned
- Maintain control of the premises
- Determine whether evacuation procedures are to be placed in operation
- Take care of the injured

- Locate and use applicable emergency equipment
- Secure the perimeter
- Determine what management personnel are to be contacted in what type of emergency, as well as when and how they are to be contacted
- Communicate all information from the scene to the command post upon request or in a timely manner
- Monitor critical areas for safety
- Continue to respond to calls for service
- Coordinate traffic control around the facility; assign someone to direct public safety responders to a designated area and maintain an open flow of traffic onto the facility grounds
- Coordinate and cooperate with all public safety authorities
- Maintain company functions, if possible, in nonaffected areas
- Route the media and family members to controlled designated areas outside of the perimeter
- Determine (a) who will announce the "all clear" signal for all personnel to return to their regular duties and (b) when such an announcement can be made

In addition:

- The security or safety officer will be the initial emergency responder for first aid, fire, explosion, etc.
- As the initial responder to an emergency occurrence, the security safety officer should be trained in at least his or her assignment and responsibility, basic first aid and cardiopulmonary resuscitation (CPR), basic firefighting techniques and procedures, and how to respond and react to various emergencies.
- The security safety officer must realize that backup assistance, at least at the outset, will be minimal.
- The security safety officer at the scene will be the person to initially request outside public safety assistance: police, fire, ambulance, etc.

Chapter 22

Disaster Management of the Incident

Identification and Evaluation

How does the management of a facility or institution quantify a risk, whatever the nature, so as to be prepared for any disaster or catastrophe that could occur? This is a particularly challenging task, given that the threat of the risk is unknown as to type, time, and place. However, to assume that risks can be discounted because of these unknown factors that cannot be established is like an ostrich placing his head into the sand. But that, in essence, is what has happened since 9/11. In many respects, the security efforts to reduce the probability of a recurrence of such an event in the United States and other countries are lacking. Specifically, our railroads and ports have received little to no assistance, financial or otherwise, from our governmental institutions.

One only has to look at recent disasters such as Hurricanes Katrina and Andrew to realize the failings of the Department of Homeland Security and FEMA. Security and terrorism experts generally agree that 9/11 was only an introduction into further terrorist acts against the United States and the Western world.

The insurance industry makes use of a system known as *risk versus uncertainty,* and premiums are based on the experience of both ratings. These ratings are determined on the frequency of previous losses by the insured, the class of those insured, or the exposure which accounts for the estimation of risk established on theory or "a best estimate" based on varied data. If risk cannot be determined by either method, there is uncertainty in the formula, and insurance companies are reluctant to underwrite any risk that cannot be estimated. We have seen the recent termination of a certain percentage of homeowners' insurance policies in the northeastern and southeastern sections of our country in order to lower the risk. In reality, it was to be expected because of weather anomalies and the exceptional losses to the insurance companies.

Outline for Risk Determination and Evaluation

The following outline may be used as an effective plan for the loss-prevention manager in determining the concerns and procedures in preparing for a serious incident in a large business facility or institution.

Enhanced Awareness and Risk Evaluation

You know your building best. Survey your building or facility:

- Ask yourself, "How should I do it?"
- Can inadequacies be corrected?
- If necessary, can you plan around inadequacies?
- Are all threats reported to you?
- Do you keep a record of threats to your business and threats made to tenants in your building?

Surveys should include:

- HVAC air intakes, controls, zones and shutoffs
  - Are they monitored?
  - Automatic controls? Can they be reversed? Tampered?
  - Elevators (public, private, and maintenance) and escalators
  - Emergency shutdown availability? Ability to override?
  - Stairwells and air shafts
  - Clear of all encumbrances?
  - Entryways and fire and emergency exits
  - Clear of all encumbrances?
  - Basements, roofs, and exhaust and intake apparatuses
  - Secure from unauthorized access?
- Interior and exterior adequate lighting where necessary, including emergency lighting
- Gas, electric, water, steam
- Automatic startup of emergency electric generators?

- Backup for critical power systems, battery packs, etc.?
- Telecommunications: telephone, computer, and alarm systems
  - Accessible to the public or unauthorized employees? Controlled to some degree?
- Emergency communications: public address, portable radios.
  - Adequate quantity and availability?
- Monitoring of critical systems
  - Systems monitored by CCTV, alarms, sensors, contact points, and/or security officers?

Routine visual inspections by security officers on patrol must include:

- Be attentive for blocked or unlawfully locked fire exits.
- Be alert to all and any safety and security issues.
- Look for the unusual—bottles, cans, jars, pipes, spray cans, bags with fluids, packages, backpacks or valises found under suspicious circumstances.
- Make the safety committee a part of the disaster and emergency management process for analysis and informational input.

Existing Security Systems

All existing security systems must be inspected to determine their adequacy and effectiveness. If there is a failing in any area of concern, proper security and safety must be addressed and upgraded. Depending on the type and relevance to the facility, the security and safety systems and appropriate procedures will include:

- Visual routine inspections and regularly documented inspections
- Adequate and capable fire hardware and equipment
- Adequate first-aid supplies
- Security cameras, mirrors, motion sensors, and magnetic and electric contacts
- Metal detectors, x-ray machines, monitored reception desks
- Adequate safeguards and security systems for building access, computers, computer files and programs, and access to restricted internal spaces

Risk Identification

The loss-prevention department must consider the following vulnerable facilities:

- Government offices and buildings
- Financial institutions
- Historic monuments and locations
- Well-known national and international business entities and their subdivisions
- Tourist attractions
- Theaters, museums, sports arenas, or any location that draws large crowds

In addition, the loss-prevention professional must also be aware of particular business locales. For example, in the case of a high-rise building with more than one tenant, it would be incumbent to know who the other tenants are. If one or more tenants face a high risk of terrorist or other criminal activity that could affect the inhabitants or operations of your business, loss prevention should be aware of these threats and the safeguards in place for the building tenants as a whole.

Consideration must also be given to any business that has a mass-transit entrance, exit, or terminal in its building. Such a situation would require cooperation in security and safety concerns between the private sector and the civil authorities that have management and control of the transit system involved.

Training Assessment

Does everyone who will be expected to react to a certain incident know his or her roles and responsibilities? Determine the level of emergency training and certification for certain employees:

- The loss-prevention manager, security or safety manager, and fire safety director (as applicable)
  - Experience, expertise, training and/or certification required for the position
- Security officers, guards, and safety officers
  - Security and safety procedures and response techniques, security and safety investigation and inspection protocol, legal and liability issues, exposure to statutory law and civil litigation
- The fire brigade and/or fireguards (as applicable or as required by the municipality)
  - Key employees trained in basic fire science, fire fighting techniques, and use of fire fighting hardware
- First-aid responders
  - Key employees with at least basic certification in first aid and CPR; such training should also be part of the job description for security officers

Training

- Must be current and effective
- Must be conducted regularly and reinforced in a timely manner (reinstruction and drilling)
- Must be applicable to the facility and personnel
- Must be documented
- Various emergencies are to be identified with applicable personnel trained for specific duties

- As to all employees not assigned or trained as noted in the above assessment:
  - Routine review of emergency evacuation procedures, and the location of fire exits and fire stairways
  - Review of the location of fire extinguishers and how to use them, particularly in high rise buildings, the knowledge of designated areas where battery-operated portable radios, flashlights, extra batteries, and first-aid kits are located
  - Members of the computer incident response team (CIRT) who have specific expertise must be constantly aware of and be familiar with technological advances and intrusions to the company's telecommunications and computer systems

It is essential that all employees become aware of how to react to entrapment by debris if escape is blocked by wreckage, rubble, or fire. This should include how to summon help if trapped by shouting or screaming or using some implement (cellphone, a whistle, flashlight, hammer, or any metal object for signaling).

Internal

- Effective utilization of and constant review of the Emergency Procedure Plan
- Assignment of duties and responsibilities
- Location and use of all emergency equipment
- Normal and alternative communications systems
- Key areas that require essential protection
- Conflict resolution
- Constant and documented drills for all types of emergencies
- Routine training and certification of various emergency responders (first aid, CPR, fire guards, members of the fire brigade, fire wardens)

External

- Assorted training by outside public safety authorities:
  - *Police department*—bomb threats, bomb search procedures, investigative techniques, etc.
  - *Fire department*—types and means of response, evacuation procedures, fire inspections, handling and use of fire equipment, etc.
  - *Contracted certified trainers*—for training in first aid and CPR for key personnel, and training in hazmat exposure if there is a possibility of an occurrence in or around the facility

- Awareness of the fire response plan set by the company in coordination with the fire department and central station
- Awareness of hazmat response and initial duties by in-house personnel at the scene of or near the occurrence
- Awareness of personnel response to civil disturbances and criminal activity

Drills and Inspections

All drills must be considered part of the training process and, depending on the type of response required, must be routinely conducted and documented. After completion of surveys, planning, and training:

- Test your planning
- Test your training
- Test your equipment

The inspection process must be ongoing, in that all members of the loss-prevention department must visually inspect for safety and security all operations, departments, spaces, and personnel on a routine basis with appropriate documentation for future reference.

In regard to criminal and civil litigation and the seriousness of its effect upon a business, loss-prevention management must consider that the best defense against any criminal or civil liability, particularly vicarious liability, is a documented inspection process to prevent personal injury and loss of life in any business establishment.

Therefore, consider that to counteract a criminal complaint or at least minimize litigation awards, the importance of reasonable, competent, and documented training of key personnel and premises inspections for safety and security by loss-prevention or safety personnel cannot be overstated.

In essence, maintain a paper trail of all training of personnel and the routine inspection of all security and safety concerns.

Response and Control of the Incident[44]

The Responders

Assignment, Training, and Function

In order to prepare for any disaster or emergency, a responsible business entity must be prepared to respond effectively and efficiently to take control of an incident, reduce risk to human life, and reduce the risk of property damage.

The loss-prevention manager must realize that, for any serious incident he or she must have trained personnel to respond to a given incident. This must include certified first-aid and CPR responders as well as employees trained and assigned to the fire brigade. Specific duties should be delineated in the Emergency Procedure Plan and/or the Disaster Procedure Plan. First-aid responders, fire brigade, and other personnel who may be concerned with the particular incident in question must be appointed and delegated to their particular area of response. These assignments must be strictly controlled so as not to cause duplication of services and confusion.

Public safety personnel (police and fire) will depend on information given to them by the command post for the allocation of their personnel and equipment. Personnel assigned to the command post must have available communication so as to adequately control, deploy, and understand responders' location and needs.

The training of these company personnel must be detailed and routine in nature, so that when an incident does occur, the responders will act professionally and decisively.

Moreover, the company may wish to assign a particular location as a staging area for the responding emergency equipment and apparatus, with an appropriate employee assigned to direct responders to that staging area or the scene if necessary.

Assignment and Support Services

The following considerations should be addressed and be included in the Disaster Procedure Plan, which may or may not be a part of the Emergency Procedure Plan:

- The command center: location, arrangement, and inventory of supplies required
- The director or substitute who will be in control of the command center during the emergency and the coordination of all response teams and response requests
- Identification of responders: who is to be involved and in what type of occurrence?
- Identification of the disaster team leader who will be in charge at the scene
- Identification of the post supervisor and/or key management personnel for the command post; who will coordinate that position

- Identification of key support personnel: engineers and maintenance personnel, housekeepers, employees trained as EMTs or first-aid responders
- Identification of the media spokesperson
- Coordination and cooperation with local public safety authorities: police, fire, ambulance, hazmat units or services, office of emergency preparedness, utility companies (electric, gas, telephone)
- Private emergency support:
  - Telecommunication and radio companies and service
  - Computer-related services
  - Alarm and central station services for fire and burglary
  - Glass replacement, outside recovery and restoration teams, transportation services

The Response Process[45]

1. Initial observations by response personnel upon arrival at the scene:
   - Survey the scene upon arrival. Is there a fire, gas leak, or other hazardous condition that exists?
   - Determine the number of vistims. If there is more than one, are all victims in same area?
   - If applicable, notify other emergency responders of any unusual situation.
   - Request the necessary medical and/or public safety assistance—police, fire, ambulance, emergency utility services, etc.

2. Depending on the severity of the incident:
   - Emergency personnel responding to an incident will commence with the above actions.
   - Immediately deploy the command center which will:
     - Take direct control of the incident and all personnel concerned.
     - Respond to requests by emergency responders.
     - Initiate notifications and other procedures required under the Emergency Procedure Plan.
   - Initiate stabilization of the occurrence (see the following *Stabilization Process* section):
     - Assist all injured and disabled persons.
     - Verify that adequate resources are beginning to arrive or are present.
     - Assist the public safety services and coordinate free access to the scene by these services.
   - If the incident is hazardous or toxic in nature (or if it is believed to be so):
     - Secure the area.
     - Evacuate the area of all persons.

  - If there is a spill and if possible without placing self in danger, attempt to contain the liquid or contaminant until professional help arrives.
  - Identify the type of contaminant if possible and the extent of contamination.
  - Request hazmat assistance.
- Notify utilities of damage.

The Stabilization Process[46]

After a serious incident occurs, emergency activities will begin to take place. During this period of stabilization, the actions of the first responders are critical:

1. Local resources such as police and fire personnel and apparatus will respond to the area of the disaster, and in their effort to control the incident most probably will cause more confusion by blocking access to the stricken area. This result is termed *resource convergence.*
2. Because of the incipient confusion of the event, initial responders to the scene may have some difficulty in assessing the true nature of the incident.
3. Initial responders will have only limited resources to fulfill their tasks.
4. A command structure must make critical decisions, assign personnel where needed most, direct responding resources, and take control of the incident.

It is under these conditions that the establishment of a command post should be initiated and manned by competent personnel.

Considerations

Direction and control: How the command structure/command post will operate, the authority granted by company administration to direct and assign, and the authority to make critical decisions, all in an effort to stabilize the incident. Procedures must be in place for the cooperation with public safety officials who will arrive and take control of the command center. Remember that in any major emergency situation, upon the arrival of police, fire, or other public safety authority, they will have complete control over the emergency incident. Also, they will have complete control over attendance to the injured and further prevention to the loss of life and property, and the subsequent investigation, if required.

Allocation and utilization of resources: Assurance that no matter what type of an incident, the command post has the necessary resources at hand for initial response or the response of emergency services will be timely.

Communications: Effective and efficient equipment and components must be available to the command post and all responders when needed. Hand-held portable radios are most efficient for emergency communication.

Medical treatment:[47] The highest priority of initial responders is the saving of lives and aiding the injured. This is most important during the first hour of a life-threatening injury if a life is to be saved. This one-hour period is crucial, and the longer it takes to treat and stabilize a severely injured person, the lower is the chance of survival. Therefore, initial responders must have some first-aid training so that they are aware of what actions to take during this stabilization process until adequate emergency medical teams arrive.

Following that, traumatic incidents cause emotional and stress-related symptoms immediately or soon after the occurrence. These effects can occur in emergency service personnel, employees, customers, visitors, patients, and children. In fact, anyone present thereafter will be affected in some way, some more serious than others. Some of these effects can be described as auditory, visual and time distortions, tremors, tears and hysteria, nausea, and hyperventilation. People may suffer nightmares, a feeling of constant danger or memory impairment, isolation, anger, withdrawal, anxiety, and depression. In this regard, the company may wish to provide psychological briefings or treatment to all employees following the incident, and thereafter.

Post-Planning[48]

After the incident, whether it is a bomb explosion, a serious assault upon the inhabitants of the facility, or a natural disaster, the severity of death, injury, and property damage will require extraordinary measures to meet human needs, reduce the effects of that occurrence, and achieve partial or total recovery as soon as practical. If the occurrence is of such a broad nature that it affects the populace and that social systems become disorganized or dysfunctional, the response of public and emergency agencies may be slow or deficient during the initial phases or until some order is established.

During the earliest moments of a disaster, we can assume that activity and personal conduct will be chaotic. Following that, a stabilization period will take place when adequate resources arrive and take control of the emergency. This will include attending the injured, and the prevention of further loss of life. Consider also that additional loss of life and property damage may continue to occur because of the seriousness of the disaster and the time, manpower, and equipment needed for rescue.

Following any disaster or emergency occurrence, the following circumstances will most probably occur:

- Other than the immediate and subsequent medical attention to injuries suffered by employees, post traumatic stress disorders should be considered by company management, with appropriate psychological and physiological services offered. This will depend on the seriousness of the incident and how affected the subject was by the occurrence, and any injury or emotional stress suffered by that subject will be a factor that the company must face.
- Depending on the incident and the response, civil litigation most assuredly will be taken against the company for death, injury, or emotional impairment based on some alleged or perceived wrong.

Cautions

After an explosion or serious fire, it is inadvisable to enter or remain in a damaged site without extreme caution. In any event,

- Police, fire, or building safety officials will determine if a building or facility may be reoccupied.
- Once control and authorization is returned to company management, access must be strictly controlled to those key employees and contractual services required for removal, recovery, and restoration.
- After an incident that causes severe damage, be aware that fires can rekindle from hidden hot spots or smoldering remains.
- Be watchful of structural damage to the floors and roof that may have more damage than is visible and is subject to collapse.
- During the recovery period, use care in moving about structural components or items that could cause further damage or injury.
- Normally, the fire department will determine that utilities such as water, electric, and gas are either safe to use or are turned off. Do not attempt to turn on any of these services; contact the local utility for service or repair.

Recovery and Restoration of Services[49]

The actions and concern by management, loss-prevention specialists, and other personnel regarding this part of the disaster plan are vital for returning the business or institution back into operation. Consider that the loss-prevention unit will be an important addition to the following responsibilities:

1. All personnel must be advised at the same time of the "all clear." The announcement timing will be determined by management and/or public safety personnel.
2. After it is safe to enter, secure the site from further damage.

3. Depending on the extent of the occurrence and the size of the business establishment, it will be expected that the business administration will activate a team of appropriate personnel to assess the damage:
   - Estimate the structural damage.
   - Estimate the repair of the structural damage.
   - Estimate the cost to repair or replace the damage, equipment, supplies, etc.
   - Estimate the cost of off-site warehousing, if necessary.
   - Contract for such services as cleanup, restoration, repair, etc., that may be required. Explicitly note the cost, the start and completion dates, as well as other relevant concerns.
   - Determine the extent of loss to business records, ledgers, accounts, inventory, etc., and pursue timely replacement, if necessary.
4. Determine the time required for the company to be back in business, partially or totally.
5. Determine whether the affected area is capable of normal operation. Unless management determines otherwise, all areas must be able to handle employees and visitors.
6. Participate in the recovery team. This will include security, maintenance, and engineering personnel, management and supervisors, and other specific personnel.
7. Recover as much property as can be salvaged.
8. Document and photograph all property that will be written off as a tax loss or covered by insurance.
9. Document in writing the occurrence, the injuries, the damage, and all activity surrounding the incident. The extent of the detail included in the incident report will depend on the seriousness and magnitude of the event.

The company's insurance carrier should be notified as soon as practical after the occurrence. The following actions by management, or at least by loss prevention, should take place.

1. Compile a list and inventory all damaged goods, equipment, supplies, etc.
2. Take photos of all listed damaged items.
3. Check with the insurance carrier prior to contracting with any inventory, removal, repair, or restoration services.

The ANSIR Program[50]

A primary part of the analysis agenda for terrorism is the various security specifics available to public and private corporations in protecting and preserving corporate proprietary information. One area in particular is the Awareness of National

Security Issues and Response (ANSIR) program. Essentially, it is the FBI's National Security Awareness Program, in that it is the "public voice" of the FBI for espionage, counterintelligence, counterterrorism, economic espionage, cyber and physical infrastructure protection, and all national security issues.

The FBI is the lead agency for assorted national security concerns. This includes counterintelligence activity as well as theft of U.S. technology and sensitive economic data by internal and foreign intelligence services and competitors in order to compromise, collect, or destroy proprietary corporate information using various clandestine means.

Under this agenda, the FBI has designed this program to provide unclassified information about national security threats and warnings to all U.S. corporate security directors and executives, including law enforcement and government agencies. It is directed to focus on the response capability that is unique to the FBI's jurisdiction: law enforcement and counterintelligence investigations and required compliance. Moreover, a particular focal point is directed toward the "techniques of espionage" in regard to national security awareness information to industry, so that individual industry representatives can determine their own vulnerabilities.

Each of the FBI's 56 field offices has an ANSIR agent or coordinator who, as the contact person, is responsible for receiving and categorizing appropriate intelligence. Once the information has been received and it is determined that a response is required, each office is equipped to advise and provide any national threat and awareness information routinely to all recipients within their jurisdiction via e-mail and fax networks. Initially, these notifications were designed to handle as many as 25,000 individual U.S. corporations. At this time, e-mail notifications have increased the number of recipients to 100,000, thereby increasing the register to accommodate any interested U.S. corporation wishing to receive this type of information.

In summary, any loss-prevention professional, security or safety manager, or the manager or administrator of any U.S. business establishment or institution that wishes to be added to the ANSIR program in order to receive ANSIR e-mail or ANSIR-FAX, or if any company associated with critical technologies or suspects possible foreign intelligence activity, should contact the FBI ANSIR coordinator at the nearest FBI field office.

See the Glossary for further information.

National Security Threat List[51]

The mission of the FBI regarding foreign counterintelligence is determined by a strategy known as the National Security Threat List (NSTL), which includes any threat issue, regardless of the country of origin, and a classified list of foreign powers that pose strategic intelligence threats to U.S. security interests.

In summary, the key threat issues are:

Terrorism: Foreign-power-sponsored or -coordinated violent criminal acts in order to cause death, injury, damage, fear, and apprehension against the populace of a country.

Espionage: Foreign-power-sponsored or -coordinated intelligence activity involving the identification, targeting, and collection directed at U.S. national defense information.

Proliferation: Foreign-power-sponsored or -coordinated intelligence activity against the U.S. government or U.S. corporations, establishments, or persons that involves the proliferation of weapons of mass destruction (nuclear, biological, and chemical weapons and their delivery systems) or the proliferation of advanced conventional weapons.

Economic espionage: Foreign-power-sponsored or -coordinated intelligence activity against the acquisition of sensitive financial, trade, or economic policy decisions; proprietary information; or critical technologies.

Targeting the national information infrastructure: Foreign-power-sponsored or -coordinated intelligence activity against the U.S. government or U.S. corporations, establishments, or persons, and whose activities include denial or disruption of computer, cable, satellite, or telecommunications networks or services. Also the monitoring of these systems and the disclosure of proprietary or classified information stored within or communicated through these systems, and the modification, destruction, or any use to commit fraud, cause financial loss, or commit a federal crime by using any one of these methods.

Targeting the U.S. government: Targeting of government programs, information, or facilities, or the targeting of personnel of the U.S. intelligence community, U.S. foreign affairs, economic affairs community, or the U.S. defense establishment and any related activities of national preparedness.

Perception management: Manipulating information, communicating false information, or propagating deceptive information and communications deigned to distort the perception of the public, domestically or internationally, or U.S. government officials regarding U.S. policies, ranging from foreign policy to economic strategies.

Foreign intelligence activities: Foreign-power-sponsored or -coordinated intelligence activity conducted in the United States or directed against the U.S. government, its corporations, establishments, or persons not described by or included in other issue threats noted above.

Title 18, USC, Section 3071

A recently enacted amendment to the law (Title 18, USC, Section 3071) authorizes the attorney general to make payment (up to $500,000 reward) for information of espionage activity in any country that leads to:

- The arrest and conviction of any person for commission of an act of espionage against the United States
- The arrest and conviction of any person for conspiring or attempting to commit an act of espionage against the United States
- The prevention or frustration of an act of espionage against the United States

Larceny and Liability Concerns during Emergencies

Security management must be aware that, during any incident where security and facility personnel are assisting at a scene of an emergency or otherwise directed elsewhere, the opportunity by others to steal becomes greater than at any other time, particularly in a retail or hospital environment. It may be a "catch-22" situation regarding the question of which is considered more important at the time: some emergency or a theft that may occur? We can assume that the emergency, no matter how minor, will take priority over the possibility of preventing any theft. But it should be part of the training for all personnel that they should be conscious of thieves who will take advantage of this type of situation in order to commit a larceny.

Cautionary Note

Concerning the actions and response of a security officer to any emergency, no matter how insignificant: If the policy and procedures of a company, institution, or any business enterprise that employs a security officer requires that certain actions be taken or entered into by that security officer in response to a particular incident or emergency—or if the security officer is required to act within the scope of his or her employment—and the security officer does not or fails to act in a reasonable manner, civil litigation could be sought by one who has been injured, damaged, or "wronged" in some manner. Essentially, if the security officer is obligated by his employer and through his employment to act, and if in fact fails to act or acts in a wrongful manner, he and/or his employer (vicariously) could be held civilly liable, and possibly criminally liable.

The Media

Media relations will become a major component of disaster or crisis management planning. At times, the relationship of political figures, public safety jurisdictions, and the private sector can be cordial and the media can be of use in forging a reciprocal reliance. At other times, depending on the incident to be covered and how intensely they pursue information, the media will be in conflict with whoever resists their attempt to gather information.

During any serious disaster, reporters will arrive at the scene to report on the occurrence. The more serious the incident, the more extreme will be the media response. The media have the right to respond to any incident and to honestly report the conditions that exist, the emergency response, and any other facts pertaining to the occurrence. Besides interviewing public safety personnel, the media will naturally attempt to gain comment from managerial and security personnel, since they will likely have more knowledge of the incident than anyone else.

The news media expect that reporting of any emergency relates to public awareness or the public's right to know, and any hindrance may be considered a violation of their First Amendment rights (freedom of the press) and their right to free access. Oftentimes, reporters may become abusive, demanding access and cooperation from everyone in their attempts to get information. Remember that a company and its management have a right to protect its property and prevent liability. For the sake of safety and security, access to certain areas of the facility may be restricted or secured from the media. This in particular would include a crime scene that must be protected or a dangerous area where personal safety is a concern.

Access to employee witnesses should be restricted until law enforcement authorities have completed their interviews and investigations. Comments and interviews by company personnel must issue from one person. The company should designate a manager or a responsible person who will act as a press information officer, public communications representative, or public relations authority figure and relay any pertinent information the media may request within reason. Security and other company personnel should be aware that they are not to respond to any questions by the media, referring them to the public information person for any information or comments. For very serious incidents, public safety officials will have set up a briefing area away from the disaster scene and control media access to the scene itself. Any interference by a reporter with a public safety officer (police, fire, etc.) can result in an arrest for obstruction of governmental administration. Any interference with a company responder in the performance of his or her duties on the company premises could be grounds for a charge of criminal trespass.

Remember that members of the news media are not friends and any comment will bring further invasive questioning. They, the media, are determined to get a story and, if feasible, to sensationalize it as much as possible. Any comments you make will be recorded and videotaped and, accordingly, be subject to editing or condensation to fit their report or their viewpoint. Do not place yourself or your company in jeopardy by interacting with the media unless you are authorized to do so.

SECTION 5

Chapter 23

Terrorism and Violence

Control and Prevention

A violent act may be considered terrorist-like in its nature and should be conceived as a serious threat. Although not usually classified along with workplace violence, a terrorist act is, in fact, a violent act in the workplace. The terrorist may have many reasons to commit an act of terror. When we think of the usual terrorist, we think of a group that wishes to create havoc, panic, and fear among those whom they target for an attack. The reasons could be for political, religious, social, cultural, or ethnic issues. However, the terrorist act need not be for these reasons, although the actions and the devices used may be similar. The terrorist may be a disgruntled employee or a former employee who, for various reasons, wishes to strike back at the business establishment or a customer or client and who believes he or she has been treated poorly or unfairly in some manner. Any serious assault that occurs on business premises can have devastating effects. Although we may never know in advance when a deranged or troubled individual will enter a place of business and commit a horrendous act, we can rely on certain criteria to identify an employee who may have tendencies toward unlawful behavior.

Damage, death, and injury by fire—whether accidental or intentional—are more invasive, aggressive, and costly than the average person believes. So, too, damage by intentional explosion cannot be foreseen without prior intelligence; such damage is therefore hard to prevent, no matter who commits the act. If a bomb threat is made at a known location, evacuation will, of course, minimize injury and destruction. In any event, the devastation following the use of an explosive or incendiary device that can cause serious injury, death, and damage cannot be overstated, particularly if the facility in question is well known, such as an international company, a governmental agency, a landmark, or an institution of eminence.

The loss-prevention or safety manager must be constantly aware that some minor act or incident could transform into a major circumstance within moments, and death, injury, and damage could escalate to dreadful proportions. Our objective at this point is to cover certain incidents in particular that can seriously affect any business: assaults and violence in the workplace.

Chapter 24

Violence in the Workplace

Bearing in mind that violence in the workplace may appear to have nothing to do with disasters, one only has to think of the incidents that have occurred in the recent past. Assaults in varying degrees have occurred in post offices, schools and universities, institutions, and other work locations. The offender might be an angry workman, a mentally disturbed student, or an enraged spouse. Whatever the occurrence or the reason, death, injury, and destruction of property can be excessive. As an afterthought to what may have already occurred, awareness of the causation of the incident, along with controls and preventive measures, could have at least reduced the seriousness or results of the act.

According to the Bureau of Labor Statistics (BLS) National Census of Fatal Occupational Injuries (CFOI), criminal homicide is the second leading cause of occupational death in the United States. Nearly 1,000 workers are murdered and 1.5 million are assaulted in the workplace each year. Approximately 80% of the homicides are shootings and stabbings. Not all homicides and assaults are perpetrated by coworkers, but the news media tend to sensationalize acts of workplace violence committed by coworkers. In fact, the most common motive for job-related homicide is robbery, accounting for 85% of workplace deaths.

Based on studies by the National Institute for Occupational Safety and Health (NIOSH), risks of death and assault are greater in some industries. Taxicab operators suffer the highest risk, followed by retail sales clerks, the police, sheriffs, gas station workers, and security guards, in that order. In fact, the National Crime Victimization Survey (NCVS) determined that retail sales workers were the most numerous victims, accounting for an average of 330,000 attacked each year. While these studies appear to emphasize violence among workers, the fact cannot be ignored that violence is more likely to come from outside the workplace.

People expect a safe environment, whether it be a business establishment, a stadium, a mall, or a hospital, to name a few. If some harm befalls them purposely or by chance, criminal and civil liability will certainly ensue against the perpetrator, and vicariously to the facility and its staff, particularly those in security and safety. Naturally, the more serious the incident, the more liability will be established by counsel on the plaintiff's behalf. The end result is that the company will ultimately suffer losses economically, commercially, in good will, and in productivity.

Workplace Violence Defined

The most common understanding of workplace violence includes disputes or confrontations between employee and employer that escalate into serious altercations, and where serious injury or death may result. Such actions can occur in and around business premises. However, as we have seen in recent years, violence can be precipitated by other factors that do not include the employee/employer relationship. The increase in workplace violence has risen so rapidly that no organization can be sure that it is immune.

Violence in the workplace is not only a substantial contributor to occupational injury and death to employees, but also to the threat of personal assault on the general public who may visit or inhabit the place of occurrence. Other various factors will come into play, depending on the gravity of the incident and may include property destruction, business disruption, employee absence, and loss of company good will. Moreover, the question of civil liability will surely arise and considerable monetary penalties may seriously affect the business enterprise.

OSHA Categories of Violence

Although there are no current OSHA (Office of Safety and Health Administration) regulations that apply specifically to occupational homicide or violent acts committed within the workplace, it is mandated in OSHA's General Duty Clause that an employer must provide "a safe and healthful work environment for all workers."[52] This section of law delineates OSHA's commitment to encourage employers to develop workplace violence prevention programs. The employer may be subject to fines if it fails to do so. Million-dollar lawsuits for wrongful death, injury, or negligence could be based on this standard.

Workplace violence can be defined as any incident in which one or more employees are abused, threatened, or assaulted in an incident arising in the course of employment. The violent act may also be inflicted on any occupant other than an employee. Offenders can include a coworker, former coworker, student, former student, family member, spouse or estranged lover, customer, client, patient, vendor, or a member of the general public.

As described by the Office of Safety and Health Administration (OSHA), violence in the workplace falls into three categories:[53]

Type I. Stranger Violence: This type of violence occurs when a stranger, unknown to the occupants of the premises—business, school, institution, or place of worship—enters with the intent to commit a crime. The situation could include an armed robbery, in itself a terrifying predicament without a shot being fired, or some other overt action taken by the perpetrator. It could include a sexual assault upon an employee, customer, or visitor, or a deranged or hate-filled individual who believes that he or she has been offended and seeks satisfaction by way of murder or assault. Or the deranged individual might believe that such action will bring him or her some type of recognition, whether reasonable or not. It is estimated that this type of violence accounts for 60% of all workplace homicides.

Type II. Client Violence: This type of violence occurs when a client attacks an employee during the course of a business transaction; for example, a welfare recipient attacking a social worker, a patient striking a health care worker, or a disgruntled customer attacking a sales person. Such attacks come from an individual who has become frustrated or provoked by the denial or delay of a benefit or a service. It is estimated that 30% of all workplace homicides fall into this category.

Type III. Employee Violence: Although this category accounts for only 10% of workplace homicides, it appears to be the most prevailing aggression as perceived by the general public. This type includes employee versus employee; employee attacks upon supervisors, managers, and bosses; and domestic violence. The subjects of the conflict can also include former employees, temporary or contractual employees, or subcontractors.

The available percentage figures for these three categories are well-accepted but recent criminal activity is changing the percentages and the types of violence within each category.

Workplace Violence: A Perspective

Workplace violence can strike shopping malls, retail department stores, offices, government buildings, schools, banks, libraries, and even restaurants. The escalation of violence in these accessible public locales endangers not only the employee, but also the visitor. The nature of the assault may be random, or it may encompass targeted individuals.

Nevertheless, major clusters of violence are found in particular occupational settings. More than half (56%) of workplace homicides occur in the retail trade and service industries. Following that, finance, insurance, and real estate occupations also are leaders in workplace homicides. Between 1980 and 1992, 76.4% of work-related homicides were committed with firearms. Homicides resulting from wounds inflicted by cutting or piercing instruments totaled 12.4%, with other methods of homicide rounding out at 11.12%.[54] The National Institute for Occupational Safety and Health also reported that, in 1997, employees murdered more than 100 bosses and coworkers.[55]

The occurrence of violence that can occur in Type I and Type II categories is exceptionally hard to identify, since these assailants will most likely be strangers. Once that person enters your premises, it will usually be the first time you will have or be aware of any interaction with him or her. Other than certain safeguards that will reduce crimes such as robbery, and specific actions by an employee to reduce or curtail an attack, this threat can never be completely eliminated. Although minimal, these safeguards and actions will be discussed later in this chapter.

Therefore, there is only one category where we can apply some effort in addressing the problem—Type III, employee violence. Since there can be no definitive inquiry into a person's private life during the hiring process, the employer must depend completely on prior employment and hope that the information received is complete and accurate. Other than the initial pre-hire interview, where certain information may become known, the employer has no further recourse other than to observe the demeanor and performance of the newly hired employee during his or her probationary period. If a disturbed individual does not manifest odd behavior during this initial period, sooner or later there will be obvious and clear warning signs that can alert an observant supervisor or boss to possible serious occurrences. Although the employer may view workplace violence as a minor problem, there is also the problem of compliance under federal law that mandates the employer to provide a safe and healthful work environment for all employees.[56]

Specific Acts of Violence

Although the following types of incidents may occur in one or more of the above classifications, such occurrences are becoming more prevalent and will be addressed here in some detail.

Office Rage: Hostility in the Workplace

According to an article in *Newsday*, a 1996 Gallup nationwide phone survey of 1,000 adults, age 18 or older and employed full or part time, found that nearly 25% of those respondents indicated that they were "generally at least somewhat angry at work." Donald Gibson of Yale University and co-author of the study *The Experience of Anger at Work: Lessons from the Chronically Angry* noted that this statistic may help explain recent outbursts of office rage (workplace violence). Comment was made concerning the July 29th shooting spree in two Atlanta, Georgia, office buildings where 9 people were killed and 12 were injured by a subject who was a day trader at those locations. A week later, a subject killed three people at two Pelham, Alabama, companies where he had worked. The study, which was to be presented to The Academy of Management at an annual meeting on August 11, 1999, notes, "There is a sort of undercurrent of anger and resentment aimed at the workplace that could potentially lead to the kind of explosions of rage we have seen." An interesting comment found in this study was that the most common cause of workplace anger, cited at 11%, was the actions of supervisors and managers.[57]

Domestic Violence

This is another area of violence that appears to occur in the Type III category that should be of concern to the loss-prevention manager. Domestic violence results from personal relationships. Spouses, estranged domestic partners, ex-lovers, and others affected by romantic obsessive behavior may seek out their victims at the work site to initiate or continue an ongoing argument. Many times the incident may only entail a loud argument, but if an altercation ensues, an assault will usually take place. Because of embarrassment, the involved employee rarely makes known to management the circumstances of the domestic problems. However, other coworkers may be aware, since they may have seen the results of a beating or observed the emotional state of the victim. The victims of such violence must feel secure in advising their employers about the risk or potential of such an attack, or of an incident that may take place around the business premises. Solutions may include restriction of entry to a family member or the subject in question, a change of workplace assignment, transfer, flexibility of work hours or leaves of absence, and most importantly, awareness by security personnel.[58]

School Violence

School facilities are not immune to acts of violence occurring on or around the school grounds or at school related activities. Harassment among children, threats, minor assaults, and larceny have always occurred, but in recent years, serious acts of violence such as vandalism, rape, felony assault, and murder have occurred with all too common frequency. Even the well-to-do suburban school has suffered from these events.

A study of crimes in public schools during the 1996–1997 school year revealed that more than half of the schools responding to the survey experienced a variety of crimes, ranging from minor to very serious offenses. One in ten schools reported incidents of suicide, rape or sexual assault, physical attacks or fights upon teachers and other students, and serious assaults with a weapon. The serious violent crimes appear to occur in schools with large student populations. One-third of the schools with 1,000 or more students experienced serious violent crimes.[59]

In recent years, we have seen the homicide of students, particularly by other students at schools in all types of of neighborhoods. Note the following major occurrences just within the last decade:

1. Columbine High School, Jefferson County, Colorado; 4/20/99; 15 dead, 24 injured; perpetrators: two students, age 17 and 18
2. Red Lake High School, Red Lake, Minnesota; 3/21/05; 10 dead, 12+ injured; perpetrator: student, age 16
3. Amish Primary School, Lancaster County, Pennsylvania; 10/2/06; 6 dead, 5 injured; perpetrator: adult male
4. Virginia Polytechnic Institute and State University, Blacksburg, Virginia; 4/16/07; 33 dead, 25 injured, perpetrator: student, age 23
5. Crandon High School, Crandon, Wisconsin; 10/7/07; 7 dead, 1 injured; perpetrator: deputy sheriff, age 20 years

Reasons for the assaults include easy availability of firearms, hard rock music, aggressive video games, mental disease, and harassment by other students because of odd behavior or dress. Whatever the reasons for these actions against others, this is a societal issue and not a topic for security other than awareness of possible prohibited conduct and the protection of the student population.

Although attitudes toward security in schools have begun to change, fewer than 15 years ago school authorities rarely looked upon the accessibility to the school building as a serious threat to the occupants. The absence of a serious threat caused schools to assign teachers as hall monitors, who also supervised students in cafeterias and study halls, and policed students cutting classes or smoking in the restrooms. Particularly in the urban areas and the larger schools, as incidents became more prevalent, schools began to hire security officers to patrol the buildings and grounds, and to respond to any criminal act.

Students are not to only persons to perform violent acts. Unfortunately, teachers and others in authority have committed assaults and serious sexual crimes on students and must be treated severely according to the law.

Effect of Violence on the Workplace

The full effect of a business loss may not be immediate in many cases. The initial incident may cause severe harm, damage, and an interruption of business operations, but following that, the final assessment may total millions of dollars that could have a major effect on the viability of the enterprise. Business losses suffered because of workplace violence may arise from:

- Death and injury to employees
- Death and injury to visitors, customers, patients, clients, students, etc.
- Property damage
- Business disruption, full or partial
- Civil litigation against employees or the company
- Criminal litigation against employees or the company
- Legal expenses
- Loss of productivity or workdays
- Loss of good will; negative company image

Finally, do not forget that the emotional trauma of being involved in or witnessing a violent act can be long lasting.

Causes of Violence[60]

If we don't understand the causes of violence, how can we address the problem in an attempt to protect the innocent? In actuality, violence in our society may result from economic, societal, organizational, and psychological considerations.

Economic: A distressed economy creates an overstressed population as a result of downsizing and substantial layoffs; massive mergers, with substantial job losses; growth of technology resulting in outmoded work skills and subsequent high unemployment, recession, and poor salaries and benefits.

Societal: A changing society, with a decline of shared values, reflects the effect of violence on television and in the movies, the accessibility of firearms, intrusive music, and violence as an accepted means of solving a problem.

Organizational: Dysfunctional management styles and organizational cultures (authoritarian, autocratic, or abusive) foster polarization among employees and managers, discourage new ideas and creativity, ignore the voices of employees in the organization, and fail to address grievances, all leading to poor job satisfaction. In addition, the organization may not have clearly defined rules of conduct; may tolerate negligent hiring and retention practices; may be guilty of poor training and supervision of employees; and may not be fair, firm, consistent, predictable, and timely in dealing with disciplinary issues. The organization may also maintain an atmosphere of intolerance, bigotry, disrespect, harassment, labor/management antagonism, and unresolved workplace issues. In such an environment, an employee may perceive unfair treatment, injustice, or malice in the event of termination, downsizing, or layoffs. Employees who feel that they have been treated unfairly may become disgruntled or even enraged.

Psychological: Employees bring along their "baggage" into the workplace, which affects their personality, character, and job performance. Such baggage may be the result of emotional, physical, or sexual abuse from childhood or as an adult. Stress and mental illness are the major forms of psychological dysfunction.

Stress is the body's response to high levels of anxiety. There is good stress and bad stress, and it can be acute or chronic. It is said that people need some level of stress to motivate them; otherwise, they will stagnate. Good stress may result from competition and reasonable deadlines for workers. The level of anxiety caused by good stress is in the low-to-moderate range and is acceptable for short periods of time. If the levels of anxiety last for an extended period of time, then stress becomes bad and the individual's functioning becomes impaired or overwhelmed and behavior may become noticeably abnormal. Depression or mood changes, which may be minor at first, could lead to exceptional or outwardly violent behavior. Excessive stress is the cause of most violence in the workplace.

Mental illness is another cause of abnormal behavior, which may or may not lead to violence. People afflicted with psychological illnesses may become violent under stress, but the outcome is usually more benign, resulting in distress, depression, or disturbed demeanor, as evidenced by a deterioration in social or occupational life.

Characteristics of Persons Who Commit Acts of Violence in the Workplace[61]

Any of the following attributes or temperaments may appear in a person who commits a violent act. Idiosyncrasies or odd behavior may be multiple in nature, and thus they may not be as perceptible to the ordinary person in determining what may be abnormal or an apparent threat to others. Consider the following as guidelines to the observation of aberrant behavior among employees.

History of Violence

- Criminal activity
- Physical confrontations
- Domestic or spousal violence
- Sexual abuse
- Child abuse
- Verbal abuse
- Antisocial behavior

Psychoses

- An inability to evaluate the external world objectively
- An inability to distinguish the external world from inner experiences
- Impaired reality testing
- Senseless violence
- Grossly disorganized behavior
- Partial or total inattentiveness to the surrounding environment

Under this category, the following conditions may be apparent:

Schizophrenia: Acute psychosis; may include bizarre behavior, hallucinations, poor insight (i.e., subjects may think that someone is controlling them or speaking to them)

Major affective disorders: Often present with delusions and hallucinations, and also associated with moods or mood swings (i.e., depression, delusions)

Paranoid states: A disorder characterized by delusions of grandeur and persecution, suspiciousness, jealousy and resentment

Personality disorders: Traits that become disorders; includes individuals who are very rigid or cannot be reasoned with, and the pathological liar or the blamer who will accept no responsibility for his or her actions; such disorders also include passive-aggressive behavior, where the individual indirectly resists authority, responsibility, and obligations, and whose aggressive impulses are passive in nature. (However, the antisocial personality and the person who exhibits aggressive behavior will fail to conform to social norms and lawful behavior, leading to grounds for arrest. Moreover, deceitfulness, reckless disregard for the safety of self or others, lack of remorse, and indifference as to how his or her actions affect others are generally conspicuous traits in personality disorders. But most importantly, the person who exhibits irritability and aggressiveness, which will be indicated by repeated physical fights or assaults or threats and gestures, could cause harm to another.)

Impaired neurological functioning: Individuals who have been diagnosed as having hyperactivity as a child, brain injuries, or abnormal electroencephelograms (EEG) tend to be less capable of inhibiting themselves; such behavior is to be considered only a characteristic, and not a cause of violent action

Romantic Obsession

The subject in question has an infatuation:

- The subject of the fixation may not be aware of the degree of the attraction.

- The subject of the attraction usually holds a higher social status or an important position (i.e., a celebrity, politician, or a business associate in a higher or more important position).

Chemical Dependence

The individual uses alcohol and/or drugs, on or off the job.

- Alcohol abuse can cause acute and chronic physical, antisocial, and abusive behavior.
- Illegal use of controlled substances or the misuse of prescription drugs may induce erratic or abnormal behavior, excessive mood swings, or push a subject "over the edge."

Depression

Depression may not be immediately observed, or it may be hidden or controlled by the subject, but the subject may be deeply affected by this disease.

- Most commonly treated by prescription drugs and/or counseling.
- One in seven people suffering from depression will commit a violent act directed at himself or others.

Elevated Frustration

The individual demonstrates irritation with the following:

- Environment: the work area or conditions of work, the home, the neighborhood, or his or her vehicle
- Family, peers, or coworkers
- Local, national, or international issues or conditions

Interest or Obsession with Weapons or Rebellious Groups

This area does not include hunters or gun hobbyists, but does encompass:

- Affiliation with subversive, racist, antigovernment, and fringe organizations
- An individual who constantly talks of guns and weapons
- An individual who expresses a desire to shoot or kill someone else
- An individual who hides a firearm or other harmful weapon in his or her vehicle or carries a weapon at work (i.e., a hunting or large knife or an illegal concealable firearm or other weapon)

Humiliation

An individual who has been embarrassed in some manner, particularly by a public humiliation, may consider it a profound occurrence and a cause of an increase in anxiety, anger, and depression, thereby triggering violent behavior. These occurrences may include:

- Public reprimand
- Public teasing or harassment
- Failure to receive a promotion or a raise in salary
- A critical mistake that negatively affects the company
- Termination for cause or because of company downsizing

Abnormal Behavior

Defined

- ***Abnormal behavior***: Behavior representing a change from the norm. These changes are caused by physical and/or chemical imbalances or events, known or unknown. The behavior can be acute or chronic.
- ***Normal behavior***: Each person has his or her own level of normalcy. While one person may always smile, appear happy, or outgoing, another may always appear grouchy, shy, or backward. Normal behavior is what is consistent for a particular person. Deviation from this pattern of behavior can be considered abnormal.

Definitive Observations

The following patterns of behavior are easily observed:

Alcohol abuse: This is usually easy to observe: the smell of alcohol on breath, the body giving off an odor of alcohol, bloodshot eyes, impaired ability of locomotion, and slurred speech. While under the influence of alcohol, the subject may become loud, aggressive, and easily irritated.

Drug abuse: These signs are more difficult to pinpoint, but the physical signs are similar to alcohol abuse without the smell of alcohol. Additionally, the subject may also appear to be indifferent and lethargic and may disappear from the workplace for periods of time.

Impaired judgment: The individual experiences mood swings, ability or inability to reason changes, and his or her decisions become questionable. May project mixed messages, talking positively but behaving negatively.

Emotional demeanor: The individual becomes very sad, may cry at inappropriate times, may carry on hysterically over a minor incident, or may laugh or become overly giddy unsuitable to the current situation. He or she argues frequently and is belligerent toward others, possibly yelling, shouting, and slamming doors. The individual may be a constant complainer, consistently negative, or inappropriately angry for little reason. The person may also have disruptive outbursts or temper flare-ups disproportionate to the irritation, has poor social skills, can be remote, and can become fixated on an idea or an individual.

Financial problems: The individual constantly prods coworkers for loans or employers for salary advances. Seems to be broke all the time, or always complaining of not having enough money. May use as an excuse, real or not, overdue rent or mortgage payments, car payments, family health problems, marriage problems, etc. May have serious gambling problems.

Family relations: Serious cause for abnormal behavior in an otherwise valued employee. Subject may have marital problems such as separation, estrangement or divorce, spousal abuse, trouble with children, or family illnesses.

Legal problems: Whether civil or criminal, legal problems may become known to coworkers or employers by gossip, the media, or admission of the individual involved.

Occupational responsibilities: The initial sign is an increase in absenteeism, arriving for work late or leaving work early. The individual will become less productive, with work output becoming poor or sloppy; force others to take up his or her slack; disappear from his or her workstation and take long lunch breaks. The employee may be very rigid, inflexible, and controlling in work habits. Has contempt for new technology and may withhold important information.

Threats: Threats may be observable in many forms, tangible or verbal. An individual may have a tendency to make coworkers uncomfortable and/or fearful. He or she may make threatening gestures and his or her personality may be controlling, demeaning, and demanding. Other than the actual violent act, a threat that causes fear and alarm in another must be considered significant in the curtailment of possible violence in the workplace. Any overt act, incident, or observation regarding a threat requires immediate or almost immediate action. The company must provide a procedure for reporting any type of threat, and all threats should be taken seriously.

Evaluating the Threat

Risk Factors

The following components, though not necessarily in order of risk, are serious factors regarding risk in certain occupations:[62]

- Contact with the public in the service sectors, such as community-based settings, public and customer relations, salespersons, health care, and public service personnel
- Working or dealing with unstable or volatile persons in hospitals, health care, social service, or criminal justice settings
- Exchange of money, which includes check-cashing services, cashiers, cash transport
- Delivery of passengers, goods, or services, including taxi and livery drivers, truckers, and delivery services
- A mobile workplace such as a taxicab or police cruiser
- Working alone or in small operations
- Working late at night or during early morning hours
- Working in high-crime areas
- The safeguarding of valuable property or possessions; includes armed guards, security officers, couriers, transport of money or valuables

Environmental Design

Efficient environmental design and procedures within a facility can deter or at least lessen the threat of an assault upon an employee. Such design factors might include:

- The implementation of locked drop safes, secure cash rooms, and holdup alarms
- Maintaining small amounts of cash on hand or in the cash register
- Adequate closed-circuit TV (CCTV)
- The posting of signage and notices warning of limited cash on hand and the presence of surveillance cameras
- Secured booths or enclosures (possibly bulletproof) for employees in high-crime areas or during night hours
- More use of credit cards or debit card transactions
- The use of personal protective equipment such as body armor by public safety personnel and protection from toxic substances such as pepper spray to lessen the effect of an attack
- The use of physical barriers such as the height and depth of desks and counters, half walls, or other enclosures to separate the employee from the visitor, client, or customer to reduce the risk of injury (But we must also realize that any safety device of this nature may increase the frustration of the customer, client, or patient, and therefore may be self-defeating.[63])

Increasing the Level of Safety in the Workplace

Administrative Controls

Work practices and staffing during the opening and closing procedures of the business establishment and during money drops, screening of persons entering the workplace by security officers and receptionists, increasing the number of personnel during certain hours, and escorting of visitors or patients within or between certain spaces may be considered appropriate controls to ensure internal safety.

Written policies and procedures must be implemented to encourage employees to report threats or violent occurrences, which are then assessed to suggest, evaluate, or improve safety recommendations. Such reports can help the employer to assess whether appropriate measures are present or needed, to defuse potentially volatile situations, to refine the processes for termination, and to recommend medical or psychological assistance for those employees who have threatened or committed violent acts.

Training and education must be offered to the employees. This will include prevention tactics, such as conflict resolution and nonviolent response, in an effort to defuse volatile situations that may escalate into serious altercations. Moreover, all employees must become aware of risks that may occur within the workplace.

There are four areas that should be applied and made part of the hiring, training, probation, and supervision processes in order to reduce the probability of safety and security problems involving employees at a later date.

Preemployment Screening

In an effort to weed out potential problems, the preemployment screening process is most effective.

- Have the prospective employee complete an employment history and background form.
- Train all interviewers of prospective employees to be skilled and thorough in their questioning and observation techniques.
- Check all prior work references as completely as possible.
- Check all educational and personal references.
- Conduct a criminal history and driving records check.
- Conduct a financial and credit history check if applicable.

Orientation and Training Programs

Other than the required routine safety training classes for all new personnel, the following concerns regarding behavior in the workplace must be considered.

New Hires

- The new employee must be made aware of all rules, regulations, and restrictions that apply to employment.
- The new employee should be advised of a set probationary period during which his or her job performance and suitability to the organization can be determined.

During the Probationary Period

- Make all employees aware of the causes and the effects of stress, how to identify stress factors, and the methods for handling stress.
- Provide training and promote awareness in identifying workplace violence, dangerous trends, and significant threat potential.
- Provide training in effective communication.
- Provide training in effective conflict resolution.
- Promote awareness of employee assistance programs or other counseling programs.
- Provide intensive and specific training for loss-prevention/security officers and members of the threat assessment team in identifying, responding to, and investigating violent incidents.

All Employees

Regardless of the level of risk, all employees should be trained routinely in the issues noted above in addition to the following:

- Recognizing the potential for violence
- Recognizing that constant quarrels as well as aggressive and threatening behavior are forms of harassment that affect morale, productivity, and the well-being of the employee
- The policies and procedures set up by management to control violence or risk to employees; specify the company's enforcement of a zero-tolerance policy
- Training employees on awareness, avoidance, and response in robbery, rape, and assault incidents
- How and when to report threats and violent incidents; how the company will respond to such reports
- The availability and proper use of security equipment and systems
- The proper response by management and loss prevention to any emergency occurrence, violent act, or hostage situation
- Travel safety, parking field safety, and money transport

- Detailed descriptions of the employee assistance plan, the threat assessment team, and the loss-prevention department to include the makeup, functions, duties, and responsibilities of each body

The Security Supervisor

Although employees are more aware of what other coworkers are doing and how they behave in the workplace, the diligent supervisor will sooner or later become aware of the failings or aberrant behavior of a subordinate and must take the necessary corrective action to change that employee's work habits or personal demeanor. In this regard, the supervisor should consider the following:

- Be aware that supervisors are much more likely to be targets of workplace violence by employees because of their control of discipline, evaluations, promotion, and the possible misuse of management principles in general.
- Be aware of idle threats or specific actions made to you.
- Become conscious of uneasy or "gut" feelings about an employee.
- Do not moralize; restrict criticism to job performance, interpersonal/customer relations, or attendance.
- Do not try to diagnose the problem.
- Do not be misled by sympathy-evoking tactics or excuses.
- Do not discuss drinking unless it occurs on the job or affects job performance, but...
- Remember that chemical dependence is a progressive disease and will always get worse without treatment.
- Make it clear to the employee that the organization is concerned with job performance, and, unless improved, the job is in jeopardy.
- Emphasize, if applicable, that the organization does offer assistance programs and that they are strictly confidential.
- Explain that the employee alone must decide whether or not to seek assistance.
- Accept that the problem and signs will not go away. You will have to deal with the problem sooner or later.
- Document all incidents that are important enough to require a reaction by management.

The Security Officer

Consider that the role of a security officer is one of the more important controls concerning workplace violence.

There is little doubt that violence in the workplace is on the increase. Security officers will usually be the first responders to an incident, and often must become physically involved to control the situation. The first and most important step in anticipating workplace violence is to know your employees. Hopefully the human resources department has gone beyond a cursory check of prospective employees during the hiring process so that those persons who have troubled or violent backgrounds can be identified and refused employment.

Whether precipitated by an employee, customer, client, or visitor, a violent act can occur under various conditions. Anger, frustrations, competitiveness, theft by coworkers, and stress can make tempers escalate into physical confrontations. Moreover, minor problems can increase to where harm most likely could occur. A security officer must have his or her eyes and ears tuned to the employee community, where fellowship and silence are most often the norms.

Petty instances can lead to serious overt actions; therefore, the security officer must be aware of everything going on in the workplace. Unless the occurrence has already become critical and action must be taken immediately, it is in the company's best interest for the officer to bring these minor instances to the attention of the human resources manager as soon as he or she is made aware. Theft of personal property by one coworker is a cause for concern by the loss-prevention department, and whether the culprit is identified or not can lead to physical violence. There will also be times when tempers flare and fists fly for no other reason than jealousy, work habits, or girlfriend/boyfriend situations.

In addition, certain business situations or events may trigger a violent incident among workers. These could include negative performance ratings, disciplinary actions, terminations, strikes, downsizing, love triangles, and personality dislikes. Depending on the seriousness of the situation, some immediate action may be needed. This could include intervention, perhaps consultation by managers, or immediate dismissal. In most occurrences, a warning and subsequent close scrutiny of the combatants will suffice. In any event, written documentation must be compiled on the occurrence.

But in the more critical events such as a strike, the security officer must remember that even the most docile employee will react to mob behavior simply because his or her livelihood may be at stake. During these serious situations, loss prevention must protect those employees and management personnel still on the job and the facility itself from any harm or risk. When an employee is threatened with bodily harm and an assault takes place within the facility, loss prevention should attempt to have the threatened or assaulted employee file a police report of the incident.

During all terminations for cause, a security officer should be present or in close proximity in case the situation becomes loud, abusive, or physical. The terminated employee should be escorted to his or her desk or locker for retrieval of company property or to gather his or her own private property if necessary, and immediately escorted off premises. He or she should be given a trespass warning if the company feels that it is a necessity, written if possible, noting that he or she is not to enter into any area of the business property other than that which is open to the public. Further, the ex-employee is not to contact former fellow coworkers while they are working and disrupt them while at work. The individual should be advised that any violation of this trespass notice could result in an arrest. Under certain circumstances, a trespass warning may be administered to a former employee advising that he or she is not to enter upon company property at any time, and if the person does in fact return to the property, it could result in an arrest for trespass.

Caution should also be used in how and when reasoning or evidence is presented to the employee. Do not give the impression, if possible, that another employee assisted or in any way caused the termination of the fired employee. If the employee believes that another coworker is in fact involved in his or her dismissal, attempt to deny such belief. If the security officer feels that the fired employee may take some retribution against another, advise the person that, if such action is taken, he or she will be arrested and charged with the appropriate criminal offense.

Additionally, we must consider the possibility of some event that may provoke an employee so intensely as to lead that person to seriously injure or kill people who could have been the cause of his or her predicament. A random act of this type, or the possibility of a hostage situation on the premises calls for the immediate assistance of the police and emergency medical aid. The security officer should not become involved directly with the person committing the assault or the hostage taking; the officer's only duty is to protect as many employees and customers from harm as possible by removing them from the area or crime scene. Whether the subject remains on premises or is involved with a hostage, the police will be the controlling responders once they arrive, and the security officer will need only to give whatever assistance or information the police may require during the incident.

Legal Pitfalls and Responsibilities

An assault by someone other than an employee within the business premises may or may not be beyond loss prevention's control, and is not likely to be a cause for a criminal or civil action against the company. However, there are several ways in which a company or an institution may be held liable in a civil action concerning the behavior of an employee:

- ***Negligent hiring practices***: The company can be held liable if it hires an employee without a thorough background check during the hiring process and (a) the employee thereafter commits a violent act and (b) it can be shown that such an act can be attributed to prior employment or behavior.
- ***Negligent retention practices***: The company can be held liable if it has information regarding violent actions, possible violent acts, or threats by an employee, and supervisors take no appropriate action once they become aware of such occurrences.
- ***Foreseeable acts:*** The company or institution can be held liable for damages in a civil court for any security issue involving a foreseeable act. This can be justified in the sense that the owner or lessee of the premises failed to use

ordinary care to reduce or eliminate an unreasonable risk of harm created by a condition on the premises that the owner or lessee knew or should have known about so as to exercise reasonable and ordinary care (in this case an aggressive employee). *Ordinary care* may be defined as acting in a reasonable manner based on knowledge about a condition, and the risks that may be or have been caused by that condition. For more information, see Chapter 32, "Premises Liability."

- ***Immediate threats of harm:*** The Americans with Disabilities Act (ADA) protects the physically and mentally disabled from discrimination. However, such a disability is not protected if it presents an immediate threat of substantial harm. If an employee is reprimanded in some manner because of aberrant behavior, he or she could sue for discrimination by the employer under the ADA because the employee may decide that the employer should have recognized his or her problem. But the employer can make an employee's job conditional upon getting professional help. This may qualify as a *reasonable accommodation* by the employer to the employee under the law.

Policies and Procedures

One way to minimize liability is for a company to establish written policies and procedures for any incident that may occur at the business establishment. Such written policy concerning workplace violence would be made part of the Emergency Assistance Plan (EAP), which would cover all emergency incidents. By having a written emergency procedure or a safety plan as policy, the company indicates that it is conscious of and dedicated to the security and safety of all employees and occupants, and that it is attempting to use all reasonable care and attention to prevent harm, injury, or loss of property. The plan will fix responsibility, provide for notification, note those employees who should be responsible during a particular emergency, and detail the procedures to be taken prior to, during, or after that emergency.

Additionally, workplace-violence policies and procedures will prescribe that all employees be treated fairly and equally, with appropriate control methods and training in conflict along with dispute resolution, and the implementation of reporting procedures and an active employee assistance program for personnel referral.

The plan should include and describe in detail the following:

Violence Prevention

- Develop a written zero-tolerance policy containing procedures for reporting, counseling, suspension, and termination.

- Form a threat assessment team (crisis management team). This team will evaluate, designate, and authorize the appropriate company response to all threats.
- Detail the duties of this team.
- The employee assistance program (EAP) should be made part of this plan. If an incident was not too serious in nature or where the circumstances have been resolved to a great degree, the employee may receive counseling or be referred to various other treatment programs.
- Provide violence prevention programs, drug and alcohol treatment, etc.
- Develop a plan for an emergency response unit or group and for first-aid responders (particularly security officers).
- Provide trauma training and response to certain employees (including security officers).
- Increase security measures and security training regarding workplace violence.

Response to an Incident or Report

- When a report of unusual behavior, violent acts, or a possible threat of violence is made by any employee to any supervisor or manager, the loss-prevention manager, along with the human resources manager (if it is company policy), will conduct an immediate investigation into the matter.
- As soon as practical, the initial results of the investigation shall be made known to the threat assessment team (TAT), a group of managers, which may include union representatives or employees, that will deal with workplace violence only.
- The TAT will establish procedures as to the reporting, investigation, and assessment of the incident; statistical review of past incidents; and development of strategies, the area of responsibility for each member of the team, and the confidentiality concerning the incident, the investigation, and the assessment.
- If disciplinary action is to be taken, sanctions must include counseling, suspension, or termination of employment and, if applicable, aggressive criminal prosecution.

Responsibilities of the Threat Assessment Team

The team should be responsible for:

- Implementation of employee training programs on workplace violence
- Implementation of plans for responding to threats or acts of violence
- Auditing and inspection of workplace violence programs and hazard assessment and analysis (trends in assault incidents within departments, job titles, time of day, etc.)

- Maintenance of records pertaining to threats and violent incidents, confidential information on threats and violent actions, near-assault incidents, workers' compensation and medical records, police reports, violence training records, grievances, and other pertinent records
- The authority and availability to access outside health providers and agencies, legal and law enforcement authorities, and similar business enterprises that may have comparable problems

The loss-prevention manager should be an integral part of the threat assessment team and be involved in all actions by that team.

Chapter 25

The Documentation of Poor Performance or Abnormal Behavior

The Loss-Prevention Department

The initial information or complaint concerning a violent threat or incident will be initiated by the loss-prevention manager. If information comes to the attention of a security officer, or if a complaint is reported, the officer shall immediately apprise his or her superior and request further direction. Although there will be times where the human resources or personnel department may become involved at the earliest stages, the loss-prevention department must be notified as soon as practical. This involvement is important, since the possibility of criminal or civil action initiated by any workplace violence or harassment issue must be investigated thoroughly. Consequently, the loss-prevention manager must have the authority to originate the investigation, compile approve the written report, and submit it to the threat assessment team (TAT) for further action. The loss-prevention manager must notify the local law enforcement authorities of any act that is a violation of public law.

Documentation

Consider recorded documentation as an important factor for evaluation of employee performance and conduct. The documentation should be as specific as possible, focusing on job performance and any unusual behavior. This assessment should be compiled and maintained about any employee who shows a decline in performance that is job related, or any erratic or abnormal behavior that could affect others. The collection of this data can offer management a fair and impartial assessment of the employee.

Although anyone can have an "off day," recurring patterns of erratic or unusual behavior should be noted. This data should be considered apart and separate from the evaluation that is conducted routinely for promotional and wage increases. However, erratic and unusual behavior may be also made part of that evaluation. It must be understood that the supervisor or security officer who executes this documentation of specific performance and behavior is not a counselor or a judge of the employee, but is only someone who assesses work performance. The company or institution should have some type of service, such as an employee assistance program, where the employee can be offered some type of assistance if he or she has previously shown the capability of performing his or her job, and is considered a valued employee.

Documentation Specifics

Documentation should include all the specifics of a particular incident or inappropriate behavior as noted below.

Absenteeism

- Excessive use of sick leave
- Frequent absences or days off with vague, unlikely, or no excuses
- Routine lateness
- Early departures
- Frequently asks coworkers to punch a time clock for him, or gives lame excuses for not punching in or out

Personal Appearance, Attitude, Behavior

- Deficient personal hygiene, lack of cleanliness, and sloppy appearance
- Mood swings during the workday for no apparent reason
- Reeks of alcohol, eyes are droopy and/or glazed over, slurred or incoherent speech, inability to stand erect, or constantly sniffling or wiping nose
- Repeated accidents on the job

Coworker Relations

- Complaints from coworkers, clients, customers, or visitors
- Overreaction to criticism or suggestions, particularly from supervisors
- Always complaining about job conditions, other employees, or supervisors
- Makes intimidating comments or threats about other coworkers or bosses
- Carries a grudge to the extreme
- Is racially, culturally, religiously, or ethnically biased, and makes his or her views known
- Has others take over his job responsibilities
- Keeps to himself; avoids coworkers and associates
- Threatens coworkers and supervisors with immediate physical harm
- Becomes involved in an altercation with another employee, client, customer, or visitor

Job Performance

- Disappears from his post or job site for long or short periods
- Takes long lunches or breaks
- Appears preoccupied or unable to concentrate on the job
- Routinely commits errors in judgment
- Erratic or deteriorating productivity
- Fails to follow instructions

The Initial Report

For the purpose of retaining historical data on an employee's abnormal or erratic behavior, a copy of the report must be included in the employee's personnel file and maintained by the human resources department. The original report (initiated by the loss-prevention department) must be submitted to the threat assessment team and maintained along with records of investigation, evaluation, and follow-up of the incident.

In an effort to be as complete and as accurate as possible, the report documenting the incident or behavior of that employee should:

- Be recorded promptly
- Indicate the date, time, and location of the incident
- Describe the action that occurred or the behavior exhibited
- Identify all the participants involved
- Indicate other employees or witnesses present and/or merchandise, equipment, tools, or weapons involved
- Indicate specific performance standards violated
- Indicate specific rules or regulations violated

- Indicate the consequences of the action on the employee's job performance and the operation of the unit
- Indicate response to the employees' actions
- Indicate the employees' reaction to efforts to modify or correct his or her behavior
- Be objective, recording only observations or actions, not impressions
- Indicate any injury sustained by any participant of the incident, whether any crimes were committed, whether a police report was made and if any police action took place

The report must be submitted to the TAT for investigation, evaluation, recommendations, and follow-up.

Investigation and Evaluation[64]

Following the initial investigation and report by loss prevention, the threat assessment team must immediately investigate in depth all threats and incidents that involve workplace violence. This will include:

- The collection of all facts:
  - Identification of the principals who were involved: the offender and the victim as well as all participants and witnesses
  - Description of injuries sustained and by whom
  - Identification of the cause or reason for the occurrence as well as any contributing cause or causes
  - Identification of when and where the occurrence took place
- Recommendation of corrective action to be taken: suspension, probation, termination, use of the employee assistance plan, medical and/or psychological referral, or legal referral or assistance
- Consideration of changes in control, policy, procedures, and, if indicated, commence environmental and work-site evaluations
- Completion of appropriate follow-up

Record all information in the required and/or necessary reports. Be aware of laws governing confidentiality and the retention of records.

Records

The threat assessment team must record and maintain all information concerning violence on or around the business premises. These records must be kept up to date and maintained for the purpose of avoiding liability, discovery of trends, and statistical accumulation and measurement. These records should include:

OSHA 200 log: Must include (a) the mandated entries of employee injury and illness that require more than routine first aid and (b) the causes of loss of consciousness, required modified work assignment, or loss of time from work. OSHA also directs the posting of this log, and the notification of fatalities, catastrophes, or accidents involving three or more employees. (This log is usually maintained and updated by the human resources department.)

Assault incidents: Must include the date, time and location of the incident, a description of the type of assault, circumstances, weapons used, persons assaulted, perpetrator (if known), injuries sustained (if any), medical aid, police involvement, and time and/or work hours lost.

Inspection and evaluation: All reports, memoranda, and directives that contain findings, corrective actions, recommendations and control methods to be implemented, and the response engaged by the company shall be maintained by the threat assessment team. All meetings by this team will include the taking of minutes, which will be maintained chronologically or numerically as warranted. Moreover, any information noted in the minutes of the safety committee regarding workplace violence or the suspicion of such violence must be brought to the attention of the threat assessment team and acted upon.

Employee questionnaires: All employees must be canvassed routinely via written questionnaires requesting their views and comments concerning any observed security lapses, high-risk work areas, and relevant activities.

Secondary records: Any record that involves an incident where workplace violence took place must be collected and maintained by the threat assessment team. This will comprise at least a copy of the record that will be attached to and made part of the original report. These records may include employee training records concerning violence, insurance records, worker's compensation records, medical records of injured parties, and copies of any civil litigation emanating from a violent occurrence.

Remember that *all reports and records must be kept strictly confidential.* See "Establishment of Policy and Procedures Concerning Violence in the Workplace" in Appendix A for a working model of this process.

Chapter 26

Strikes[65]

There are many instances where corporations procure the services of a private security or investigative agency when they are subjected to a strike or picket situation. One of the more difficult situations that proprietary (company) security officers will face will occur when their fellow coworkers decide to conduct a strike against their employer, and more so if the security officers are members of the striking union or members of a union that sympathizes with the goals of the striking members. In order to be effective and ethical, security officers, whether proprietary or contractual, must remain neutral. They may sympathize with the strikers, but outwardly their attitude and emotions must be kept to themselves. Unfortunately, this is not always the case. For this reason, company management may wish to use outside professionals for the purpose of photographing, videotaping, and identifying the antagonists and malcontents for any subsequent criminal or civil action.

Depending on the severity of the strike and its effect on the safety and security of people and property, it will be up to the company administration to reduce its operation, to close down partly or completely, or to continue with business as usual. Remember that any enterprise has the legal right to remain open for business with as little disruption as possible, and the police will enforce that right. Corporate security's key function during a strike is to control violence and to protect people and property from harm. However, in the event that the strike may have escalated to some serious extent, the police may request that the company close down or curtail certain business operations for reasons of safety.

The striking union is protected by public law, but is also required to act within that law. Accordingly, the company and its employees (including administrators, managers, security personnel, and contractual agents) cannot commit strikebreaking tactics.

Defined

A strike may be defined as an organized collective work stoppage, usually as a last resort, to pressure an employer to accede to workers' demands. Such demands are rarely over one issue. For example, the demands may concern wages, hours worked, more or better benefits, or conditions under which the employees will work. The issue may also be a business practice conducted by the employer that should be changed or corrected. Today, we have the collective-bargaining process, which is a procedure administered by the National Labor Relations Board to allow employees to put forth their grievances or demands.

A general strike could involve the change in a political system and can include the whole labor force or most or all of the citizens of a country. Generally, a strike of this magnitude would include an agreement by a complete industry of workers acting in concert in an effort to coerce an employer or group of employers to accept their demands. There are two types of strikes:

1. ***Economic***: This type of strike can be conducted by the union for a change in wages, hours, or benefits. The union must give the company 60 days' notice of its intent to strike. This period will provide the company time to negotiate a settlement or make plans for the strike.
2. ***Unfair labor practice***: This type of strike can take place suddenly and without notice. This would include a wildcat strike or a sit-down strike in violation of a valid union contract. The company must react quickly, usually by requesting a court order to return employees to work. (However, employees who are not covered or protected by any labor union, and who do in fact conduct a wildcat or sit-down strike, do so at their peril and exposure to sanctions.)

Loyal security officers, whether proprietary or contractual, must remember that their primary responsibilities are the protection of employees, visitors, and customers from harm, and the protection of property. Some businesses are susceptible to great loss and damage during a riot or a civil unrest, which may have started as simply as a strike or a demonstration that got out of control.

A private investigative agency hired for the duration of the strike will be guided and its actions restricted by the law. An agent may be used to photograph or record the leading agitators or instigators of the strike for future court presentation, executive protection, or for the investigation (in addition to the police) of apparent sabotage to personal or company property. They are in no way to become involved in strike-breaking activities, coercion, threats, or constraints on the strikers or their leaders. Federal law, enforced by the National Labor Relations Board (NLRB), provides severe sanctions for those people, groups, or businesses that interfere with a labor union's right to strike.

As soon as an employee or manager learns that a strike will occur or has begun, the company administration and the loss-prevention manager should be advised. Subsequent to that, the company should notify the local police department, particularly if the strike has occurred or is imminent.

Picket Actions

Strikes and picket behavior are covered under various laws, and such laws determine how and under what parameters the strikers will operate. Generally, strikers and pickets may only congregate and picket off the property of the business in question and, under certain lawful conditions, may picket only on the public highway (sidewalk, curb, and street). The police will become involved, if only to monitor any violations of law and keep the peace.

If the strikers or their supporters become unruly and cause a disturbance or a business interruption, or if there is an attempt to stop or impede the flow of delivery, traffic, or people onto the business property, the police will handle the situation—not company security. Security officers must not become involved in physical confrontations with strikers unless assaulted.

The security officer, whether proprietary or a contractual agent, shall be guided by the laws of arrest and the use of force as noted in the criminal law statutes in the state in which he or she is employed.

Security and the company administration will be controlled by police actions in any situation and any request by the police during an emergency. This might include the closing down of the premises in order to protect individuals or groups from harm, prevent property damage, or to quiet or disperse an unruly assemblage.

Protection and access to the facility for deliveries, customers, visitors, and other employees who wish to continue to work during a strike are major concerns. Sabotage may also be a problem, since there will be some strikers who become frustrated as time passes and may attempt to disrupt business in various ways, some of which may be considered criminal and very serious. This could include bomb threats, the obstruction or interference of public utility services (electric, gas, and telecommunications), or outright physical damage to property. The police should be advised of any sabotage or threat of sabotage.

Picketing Issues[66]

Other than a strike and picket action by a group of employees against an employer to gain some wage increase or to gain or retain some benefit, we must consider that other demonstrations may take place that could affect a business enterprise.

Groups or crowds that may assemble to demonstrate or to picket a company because of some business practice that they feel offends them or others should be handled in the same way as a strike incident. An example of such activity could include issues such as offensive hiring practices, sexual or age discrimination or harassment practices, animal rights (retail stores that sell furs or animal products), or conduct considered abhorrent to certain religious groups (e.g., abortion clinics). If management cannot resolve the situation, the police should be requested. If the occurrence causes a business disruption or if their presence is illegal, picketers can be removed. Caution and discretion in tactics must be considered if the company hopes to avoid bad press and publicity.

Under various federal laws and sanctions, when a labor violation does in fact occur, a business may seek monetary damages, criminal sanctions, injunctive relief (judgment of unfair labor practices), and disciplinary actions against individuals or the union as a group.

However, concerning a demonstration other than a labor issue, a citizen has the right to *peaceful assembly* under the First Amendment of the U.S. Constitution. This amendment protects the right to picket, no matter whether the purpose is a labor dispute, civil rights, or other demonstrations. Generally, picketing is protected when it is for a lawful purpose, conducted in an orderly manner, and publicizes a grievance of some kind.

The following are the generally accepted rules that control and regulate walkouts and strike actions throughout the country.

The Right to Picket

1. Pickets (strikers) have the right to picket, demonstrate, and hold meetings as long as such activity does not violate local, state, or federal law.
2. Pickets need not be employees of the company. They may be other union members acting in sympathy with the striking union, or friends and family members of the strikers. However, they are subject to the same restrictions and laws governing the striking union members.
3. Pickets have the right to picket as long as it does not cause a disruption of any of the functions or objectives of the business; they may not interfere with business operations.
4. Picketing is legal as long as it does not limit or deny access of employees, customers, visitors, vehicles, deliveries, etc., to the business and any of its components. Blocking anyone or any vehicle from entering or leaving the business property, physically or by threatening behavior, is illegal. Strikers causing damage to any vehicle crossing the picket line while attempting to enter the property of the facility commit the crime of criminal mischief, reckless or criminal damage to property, or criminal tampering with intent to cause damage or substantial inconvenience. In addition, strikers causing harm to other employees or persons wishing to enter the striking premises may commit the crime of assault. If an implement is used and causes damage or injury, the criminal charge will be elevated to a higher degree. Check the local or state laws that apply to your employer for the correct statute warranted.
5. Pickets may act as individuals, but not in the name of the employer or any of its component parts.
6. Handout literature may be given out by pickets to passersby, but cannot be forced upon them.
7. Any picketing activity must be peaceful. Pickets may not jeopardize safety or the preservation of order.
8. Pickets cannot apply secondary pressure or boycotts against neutral or secondary employers or businesses.
9. The police have the authority to impose conditions and the number of pickets where they believe large groups of people are likely to cause disruptive or criminal acts.

Accepted Business Practices

What a Business Should Do

The administrators and/or management of a business enterprise may wish to engage in all or some of the following actions:

1. Upon determining that there will be some type of picketing movement against the company for any reason, company management should notify the local police precinct. The police will determine whether permits are required for assembly and/or picketing, control the size of the picket action, and regulate their conduct according to law.
2. Where picketing may be spontaneous or lack direction or organization, management may wish to inquire of the demonstrators or pickets the reason or issues for such activity against the company. If the activity cannot be resolved, the police should be called upon to examine, control, or disperse the group.
3. Depending on the number of pickets and their demeanor, police officers may or may not be permanently assigned to the demonstration. If the picketing is of a minor nature, the regular radio motor patrol car on post will give intensive patrol to the scene. Company security personnel should monitor the demonstration closely and request police assistance as may be required. If the picket line is large or must be closely supervised, police officers will be assigned to fixed posts at the picketing location.
4. Corporate management or security agents (this would include private investigators and security officers) may videotape any picketing action for the purpose of identifying any violent or unlawful act by individuals or groups (strike leaders, organizers, or strikers). Videotaping for any other reason cannot be justified and may be illegal. Check with local civil authorities.
5. Corporate management or agents may use undercover operatives or employee loyalists for the purpose of advising the business owner of any criminal acts that have occurred, that may occur, or any actions that might affect the business enterprise.
6. Corporate management may wish to proclaim a trespass advisement. Once the pickets or the organizer of the picketing action are advised and notified by business management that the picketing group, acting individually or in concert, is not to enter upon the property of the business for any reason, such intruder *may* be arrested for trespassing by company security personnel and turned over to the police for adjudication. Proper notification and documentation of the trespassing warning should be compiled for future reference and/or court action.
7. Management should advise company personnel who are not involved in the protest to avoid openly commiserating, interfering, agitating, or in any way becoming involved with demonstrators or pickets.
8. Company security personnel must be made aware of the precautions and actions noted herein; specifically, they must be able to distinguish between those actions they must avoid and those they may react to within the law and the parameters the company may authorize.
9. Most importantly, consider that whatever the size and reason for the picketing action, the media will surely be advised, and may respond to observe the strike and the participants. Company management should be ready to

respond to any questioning by the media. This will include a prepared statement ready for distribution.

What a Business Can Do

Regarding any violation by the pickets or the organizers of the picketing action that affects the business operation, causes adverse publicity, or has an effect on the goodwill of the corporation, management may seek an injunction in court requiring picketers to cease and desist. Videotapes and personal observations reduced to sworn statements may be required to bolster the initiation of any criminal or civil litigation.

1. Picketing may be limited to one site (e.g., the main entrance to the business).
2. Pickets do not have the right to picket on the business property. They should be removed to public property or public right-of-way (sidewalk, curb, street).
3. Physical obstruction, creation of a blockade, or interference with another person's rights is unlawful, whatever the protest.
4. Pickets may not block access to the business facility, its parking fields, or its property. They may not obstruct a sidewalk, driveway, parking field, or any right-of-way from use by anyone who desires to drive, walk, or in any way enter the business picketed.
5. If the number of people on the picket line appears to be excessive (mass picketing), and intimidating to people attempting to cross the line to work, deliver goods, or conduct business in any way, such action may not be considered as an attempt at *peaceful persuasion*, but may be considered a breach of the peace.
6. Picketers' autos in the company parking lot or field may be towed off premises. The business may reserve the right to park vehicles to employees, customers, visitors, and other persons who wish to conduct legitimate business.
7. Pickets may not demonstrate within a private business facility. This includes parking fields and areas owned or leased by the company.
8. Pickets may not picket on an adjacent business property without the permission of that business or landowner.
9. Pickets may not cause a disturbance or commit a disorderly act, individually or in concert. This includes loud and abusive language, obscene or foul language, offensive gestures, threats, and shoving, pushing, or fighting among themselves or others.
10. Any malicious damage to vehicles or to personal or corporate property upon entering or leaving or while on the business property must be addressed immediately with corporate and police response against the violators.
11. Trash bins or baskets may be located at staff entrances or public entrances for the purpose of discarding or disposing of handbills or handout material

from the demonstrators to employees, passersby, or visitors. Disposal must be voluntary by the individual.

12. Company personnel may remove any unauthorized postings or signage concerning the demonstration in question that are in or on the business property.

What a Business Cannot Do

An owner, administrator, or manager of a business, including its agents and employees, is restrained from certain conduct that may arise or take place regarding union activities and strikes.

1. An employer cannot (a) threaten or coerce an employee from engaging in union activities or (b) threaten to close the facility if a union comes in.
2. An employer cannot deny an employee the right to vote for union representation.
3. An employer cannot spy on union activities. This includes company informants paid for such actions or outside contractual agents (private investigators).
4. An employer cannot ask an employee about his or her union activities or attitudes.
5. An employer cannot fire, transfer, or demote in retaliation for union activities.
6. The corporate management and its agents may not interfere in any way with the picketing action, other than by lawful means (seeking a court order) or entering into some communication, interaction, and agreement in an effort to end the demonstration.

Conclusion

People have a right to protest, but such right is not unlimited. If union members wish to picket, the picketing must be lawful. In an effort to control any action that may get out of hand, the police or local governing authority may set reasonable restrictions, including the time, place, and size of the picketing group.

Chapter 27

Civil Disturbances

Labor strikes or disorders can easily turn into major civil disturbances. All that is needed is an agitator or a catalyst of some type to transform an ordinary law-abiding group into an uncontrollable mob. Loss-prevention personnel have a duty to protect company property and all the occupants within that property. If the disturbance moves onto the company's property to do damage or harm, security officers may take the necessary action to control the incident. If damage to the facility is caused by persons off the company property, security officers may not leave the property to pursue the perpetrators. Security officers are not law enforcement officers and have only the arrest powers given to any civilian. Unless the security officers have full police or peace officer powers given to them by the local municipality or the state, and personal liability insurance, they would do well to let the local police handle the situation. In this case, it would be advisable that the company and law enforcement be in communication regarding support under varied circumstances before a strike occurs.

Concerning general or major citywide disturbances such as those that occurred in 1968 after the assassination of Martin Luther King Jr. and in 1992 after the Rodney King uprising, we can see how a disturbance can evolve into a riot. These were large encounters, but even small conflicts will impair any business financially and operationally. Loss-prevention and other company personnel may have to shut down and set up barriers to prevent looters and rabble-rousers from entering onto company property. Company administration should establish a clearly defined arrest policy under these conditions. In addition, the administration may have to institute double shifts for loss-prevention personnel in order to increase the amount of trained manpower required. If law enforcement is unable to assist in circumstances such as this because their primary activity must be directed citywide, contingency plans should be contained in the Emergency Procedure Plan for the backup hiring of contractual security officers or other temporary aid if necessary.

In the aftermath of any civil disturbance, a number of criminal or civil wrongful acts may be perceived as litigious and, if so, will require legal action. In addition, company administration may initiate performance reviews and conduct critiques to determine shortcomings in performance. For these reasons, the importance of an appropriate, accurate, and detailed recording of all incidents must be made in writing and categorized by incident type. The office of loss prevention is the most likely agency to safeguard these reports, since this office will be involved to a great degree in all criminal and civil liability actions.

This review procedure will be helpful in determining loss prevention's responsibilities and functions. The initiation of a disaster committee with input from all department heads and responders can incorporate this information into a critique of the incident, which will then be forwarded to the loss-prevention or safety manager for changes or implementation, if necessary. This critique must be as objective and as constructive as possible. In particular, any deficiencies in emergency operations, staffing, and notifications must be noted, and any useful corrective measures must be established as soon as possible.

Remember that a business has the right to protect its property from any harm, damage, or destruction. Depending on the level or severity of attack by persons committing a criminal offense on or within the protected property, local and state law specifies how much force a civilian (company security) may use in protecting persons and property. The loss-prevention manager and his or her subordinates should be aware of the laws regarding arrest and the use of force.

Building Security and Access Control

If the management of the business establishment or an institution believes that the possibility may exist for a criminal or terrorist act, foreign or not, certain precautions must be instituted. The loss-prevention professional may wish to implement certain policies and procedures for the protection of all persons who enter in or on the premises: employees, visitors, and other occupants, and of course the facility itself.

Some industries give people free license and privilege to enter for the purpose of visiting, soliciting, purchasing, or browsing (e.g., retail establishments), and then there are some businesses and institutions where such license and privilege is limited. This latter group would include ballparks, museums, theaters, or any facility where a fee or other obligation, is required for entry. Restriction could also include a peripheral or body search or other precautions prior to entry. As an example, we may see conditions such as this at airports and hospitals, where entry to certain areas is restricted to only those persons who require access or must be given permission to do so.

The use of barriers and controls to restrict access to a building usually causes inconvenience. Many become exasperated that they have to be placed in the same category as a criminal or a suspected terrorist. But depending on the facility in question, an in-depth assessment of all risk factors must be made, and precautions must be fabricated accordingly. Handled correctly, professionally, and with a reasonable rationale, security precautions may be readily acceptable by the general public.

Access into a facility can be controlled by physical barriers such as fencing, metal-encased cement stanchions, large planters, gates, booths, or reception desks manned by security officers. CCTV (closed-circuit television), two-way mirrors, and magnetic identification cards—along with other types of employee, visitor, and vehicle identification—may also be used.

Access and control within the facility can be controlled by key, biometric or magnetic-card access to various spaces and rooms that are required to be locked down, with entry to be made only by authorized persons.

The following are some suggestions concerning access control by mechanical means and the use of security personnel. The initiation of such controls will depend on the business and services rendered as well as the risks that may be encountered by the inhabitants, including the possibility of a terrorist or criminal act:

- Metal detection systems to identify persons with weapons.
- Interchangeable lock, mortise and key, magnetic card key, biometric, or numeric cipher pad lock control for those areas within the building that require security and limited access to certain company personnel.
- All keys, locks, and card keys must accurately assigned and accounted for.
- The use of magnetic key cards or strict key or biometric control for access to various rooms and areas in an effort to control access to specific employees.
- The erection of reinforced cement barriers outside of a building for the purpose of stopping a car or truck bomb from exploding too close to the building.
- The use of gates, fencing, and barbed wire to control access.
- The routine patrol of security officers in the inspection of all interior and exterior spaces.
- The use of a security officer or a receptionist to control access of employees, visitors, and vehicles at an entrance to the grounds of the facility and to the facility itself.
- The use of cell phones, pagers, handheld CB radios, and alarms in and around the facility.
- The use of controlled electric door releases for valid entry.
- The use of local (exterior) alarms to control perimeter and sensitive interior doors, with electronic annunciation to a central monitor location.
- The use of motion detectors, magnetic contacts and other sensing devices protecting windows, doors, vents, and hatches not usually accessible to employees or to outsiders in particular.

- The use of CCTV, particularly in interior and exterior high-risk areas, with all video recorded and monitored by a security officer or other trained employee.
- Bulletproof glass enclosures for high-risk areas.
- The inspection and maintenance of all security equipment to ensure reliability and effectiveness.

The cost of employment of security personnel, safety equipment, and deterrents will be directly attributed to the viability and level of security and safety that the business may require. Therefore, measurements and determinations of safety and security to the various areas of the facility must be conducted and studied. Consideration should include the maximum use of present or existing controls, equipment, procedures, and manpower.

Prevention Strategies

Observation

In regard to criminal or terrorist activity, the most important quality that a security or safety officer can have is the ability to identify, analyze, and assess suspicious persons, objects, or conditions. Observations and/or examination may be conducted by staff or by CCTV monitoring.

We can't deny that the exposure of a suspicious or criminal event sometimes happens randomly or with "luck," when in fact such "lucky" occurrences happen because the officer, as a matter of normal routine, has his or her sense of observation and awareness tuned in to being cognizant of everything within sight. This trait may come from life experience, professional training and expertise, or "gut feeling."

Following is a checklist for suspicious persons and suspicious activity that could occur in any facility, particularly one that has open access to the public.

Remember, the key here is *observation.*

An unidentified individual who:

- Is observed loitering near or within a facility for an extended period of time, particularly in areas or spaces not usually occupied by nonemployees
- Appears to be wandering in areas not available to the public
- Is observed with no visible company-issued identification if required

- Asks specific questions about the facility, security-related matters, or key company personnel
- Does not respond or cannot provide a reasonable answer about his or her actions or presence when confronted or challenged by an employee
- Enters the facility wearing inappropriate clothing that does not match weather conditions (e.g., large overcoat in warm weather), which could possibly conceal a harmful object
- Enters the facility carrying a large suitcase or an oversized backpack
- Is delivering a package or other item, and is unsure as to where or how it is to be delivered or is unable to reasonably answer questions
- Is observed taking photographs, videotaping, or sketching the interior or exterior of a facility, particularly a government or an exceptionally recognized building
- Identifies himself as a public safety officer (police/firefighter), contractor, service technician, or news reporter and fails to provide bona fide identification

Suspicious activity or items:

- Two or more unidentified individuals loitering within or near the facility
- One or more individuals sitting in a vehicle for an extended period of time taking photographs or videotaping, with the overall appearance of conducting a surveillance of the facility or employees
- Someone placing a package or an object inside the building or outside the facility who then departs the area
- Groups that become uncooperative when challenged by an employee or security officer
- Any unattended package, box, backpack, attaché case, luggage, or other suspicious container found in an elevator, hallway, lobby, restroom, or stairwell; any item that appears where it would be unusual for it to be found
- Any envelope or box that has an oily leakage or visible wires or that produces a mechanical or unfamiliar sound. See "Suspicious Packages and Mailings" in Chapter 7.

Suspicious vehicles:

- Any vehicle or trailer, particularly rental vehicles, parked near the facility or at the loading dock without proper authorization
- Any vehicle illegally parked or parked at an unusual location
- Any vehicle that appears to be abandoned or has an expired inspection sticker or registration plate
- Any parked vehicle that appears to be overloaded or has contents that are leaking

- Any commercial vehicle entering the facility for delivery that does not have a proper bill of lading, or whose driver cannot or will not allow access for inspection

Effect of Building Design

Since 9/11, there has been increased interest in commercial building design and how it will affect occupant safety. The loss-prevention and safety professional must be aware of the aspects of the building design under his or her care and know how it will respond to various emergency situations or assaults. Buildings that are considered to be "high profile" related to safety concerns should be subjected to a structural vulnerability assessment by engineering professionals, who will examine potential structural damage under various conditions. This could include such scenarios as explosions—in the lobby or visitors' area, in other specific areas within the building, or on the street in front of or near the building—that would affect the building's structure.

Other incidents could include major internal or external explosions, or hostile encounters that might encompass nuclear, radiological, chemical, or biological events. Bear in mind also that natural disasters can occur at any time of the year, and depending on a building's geographical location, it may be vulnerable to one type of catastrophe more than another.

Many buildings today, even those fewer than six stories high, are constructed of steel framework with reinforced masonry material filling the spaces between the columns and beams. Those buildings with walls that do not contain reinforcement masonry materials may appear to be solid to the touch, but in reality are inferior and will respond poorly to the effects of an explosion or nuclear blast. In cases where the building lacks reinforcement, or has exterior "skin" damage, broken windows, debris, and overhead material or facades falling to the surrounding grounds can impede evacuation and rescue.

Generally, prior to 9/11, only fire was considered the most extreme cause of damage, death, and injury, and training, drilling, and evacuation had formal requirements to follow. Loss prevention must consider that planning for sheltering-in-place, in-building relocations, partial or total evacuations, drills, training, and written policy detailing implementation must be part of the company's Emergency Procedure Plan.

Protecting Vulnerable Areas

Admittance to and emergence from the workplace must be considered an important control in the protection of all occupants within the workplace. Absolute protection can be achieved to a great degree, but the amount of control required to achieve the desired level of safety may entail giving up certain practices and could have an effect on the business operation. How much access is to be given to the public by a business establishment and the control and monitoring of visitor movement will depend on the nature of the business. Restricting certain practices to achieve safety and security may cause annoyance, aggravation, and some harassment to the customer, client, or visitor and, in actuality, detract from the good will that a company wishes to project. Depending on the business, irritants of this kind may be unavoidable. It has been observed that the more a threat is present, the more understanding the public will be in accepting controls and restrictions placed upon them.

Internal Building Security

Depending on the accessibility given to the public upon entry into the facility, some businesses may find that access control may require some or all of the following procedures:

Access Control

The control of any employee within any facility, specifically those companies or institutions that must restrain access to many or all spaces should be considered a major, if not the most important, aspect of access control.

- Access to any office, work section, and sensitive or secure areas must be controlled by lock and key, key card, biometric or numeric cipher pad locks.
- All keys and card keys must be assigned and accounted for.
- Periodic PIN changes of numeric cipher locks or key cards should be completed on a routine basis, particularly when an employee has left the employ of the company for any reason. This would include the changing of locks or lock cores of those employees issued permanent keys, particularly those who had access to sensitive areas.

Employees

All employees should be issued identification badges that must be worn at all times within the facility. Badges should include the following:

- A recent photograph of the employee
- A reproduction of that employee's signature

In addition, depending on the level of security desired:

- The badge must be color-coded according to access given to that employee.
- The badge must have magnetic code encryption for access to authorized spaces.
- If required, the badge should contain the employee's fingerprint.

- The badge could also contain other personal data identifying the employee.

Visitors, Contractors, Vendors, etc.

Again, depending on the security that a facility may require, the following might be put into effect for persons other than employees:

- All visitors and contractors should identify themselves at a security or reception desk and sign a visitor's log. This desk should be located at the entrance of the facility.
- All visitors should be issued a distinctive color-coded visitors pass. All contractors should be issued other types passes identifying them as contractors.
- All passes must be valid for only the day of issuance. All passes must be worn on the outermost garment and be visible at all times while on the premises.
- Appointments of visitors should be confirmed personally or by telephone prior to allowing access.
- When required, company or security personnel should escort the visitor to his or her meeting place or, if necessary, accompany the visitor for the duration of the visit. Visitors should be discouraged from wandering unescorted within the facility and should be challenged.
- Access to sensitive and secure areas of the building must be strictly controlled, and if anyone is found under circumstances other than as required, he or she must be challenged.
- A company employee or a security officer should accompany all vendors and contractors. They should never be left unattended or allowed to roam about the facility.
- Members of this group must understand that all toolboxes, lunch boxes, packages, etc., carried into or leaving the facility are subject to inspection.

Protection from External Attack

The loss-prevention professional should identify specific areas of vulnerability and act to protect these interior areas from external attack by a subject intent on personal harm and/or destruction of property. These areas may include:

Lobbies

Lobbies that have public access are to be considered vulnerable to bomb attack. Telephones should be freestanding and wall-mounted—no phone booths. Telephone books, if for public use, should be hung on a chain and not encased in a cabinet or similar contrivance. Chairs or couches should be freestanding, without skirts or other material that may provide a hiding place for a bomb or incendiary device. Areas beneath these chairs should be open so that the area may be easily viewed. If necessary, the bottoms of chairs and couches should have a sheet of plywood nailed or screwed to the underside so that any illegal device cannot be placed and/or hidden underneath out of sight. Display cases should be of the type that can be secured to prevent easy placement of an explosive device in, around, or beneath them. Fire extinguishers should be recessed into the wall and secured by a glass door that requires a key or glass breakage for access, and the interior can be readily observed. Trash receptacles should have mechanical access control that is small enough to accept paper refuse only, or be of the wire basket type with clear plastic bags to contain trash.

Restrooms

Unless undesirable as a poor business practice, all restrooms and lavatories that may be available to the public should be kept locked to limit unrestricted access. Secured restrooms reduce the probability of covert bomb placement, personal robbery, and sexual attack. Drop ceilings should be secured to deny access. Trash receptacles should be similar to those placed in lobbies. Cabinets and equipment closets should be locked.

Storage Closets and Maintenance and Equipment Rooms

These spaces must be kept locked and secured at all times, with only authorized personnel assigned to enter therein. Of particular importance are sprinkler, telephone, electric, gas, and computer equipment rooms. All sprinkler post indicator valves (PIV) should be chained in an open position and secured with a padlock to prevent an unauthorized closing, intentional or not, prior to a fire emergency. These valves or gates that supply outside water to the internal sprinkler system are to be closed only during maintenance or repairs to that system. Keys to these padlocks should be secured and made available only to personnel who require access to this equipment. In addition, security personnel should "walk the system" routinely in order to make visual inspections of all shut-off valves at or near water pipe risers.

Unauthorized Areas and Spaces within the Facility

Postings and warnings (signage) for those areas not open to the public, but cannot be completely secured, should be easily visible. Any suspicious or unauthorized access observed by an employee should initiate a confrontation and be brought to the attention of security personnel or management immediately.

Controlled Areas and Spaces within the Facility

There should be strict control of access to areas vulnerable to internal attack by employees or other persons, who must be excluded from any area where they have no business purpose. These spaces will include:

- Money safes, computer files, records and programs, and business records safes
- Business records, business plans, and payroll and accounting areas
- Electrical, telephone, sprinkler, and other maintenance/hardware rooms or storage
- Computer equipment and computer files storage rooms

External Building Security

The following areas may be considered vulnerable to a variety of accidental, criminal, and terrorist threats:

Facility Perimeter

Depending on the assessment of the protection required for a particular facility, fence lines should be protected by CCTV remote cameras, electrified fencing (with properly posted warnings), or in-ground motion or other type movement detectors. Any intrusion must signal a central security post for immediate response. Open land areas should also be controlled by CCTV cameras and other monitoring devices in order to monitor intrusion. Shrubbery, particularly located near or next to the facility, should be removed or should be well trimmed and sparse enough for any observation of suspicious activity or a questionable object. In very sensitive areas next to the building, heavy plastic screening may be used just below the ground cover so as to deter easy placement of an explosive device in the ground. Additionally, perimeter fencing may require sophisticated measures. These may include controlled access points manned by security officers, security dogs, CCTV cameras, in-ground motion and pressure detectors, and adequate lighting.

Also take into account those perimeter openings that could cause unlawful entry. These areas could include air intakes, exhaust fans, air shafts, and all doors and windows located on the roof, basement, and the perimeter of the building. A suitable and effective burglar alarm system will include an adequate central station service along with particular areas programmed for daytime annunciation of any activated entry or door.

Vehicle Access and Parking

Access and egress, and the constant monitoring of underground or interior parking, should be strictly controlled by security personnel. Service and delivery vehicles should give advance notice if possible, with the identity of the driver, company, and purpose confirmed, along with the manifest or contents of the vehicle visually observed. Any unauthorized vehicle that attempts to enter a facility or parking area or appears to avoid security procedures must be considered a serious threat and dealt with as a serious breach of security.

A bomb contained in a vehicle can cause exceptional damage, destruction, and death, particularly if the vehicle can be parked within or as close to the building as possible. Several techniques may be employed to reduce the threat of a bomb delivered in a vehicle:

1. Placing the visitor and employee parking areas some distance from the exterior of the building
2. Prohibiting parking at or near the access points of the facility and converting those areas to drop-off zones, with continuous strict enforcement
3. Strict control of access to underground parking or delivery, with public parking in these areas prohibited and all deliveries monitored by identifying and recording vehicles and drivers
4. Placement of concrete planters or metal encased cement barriers outside of a building so as to control parking or any vehicle movement against or near the structure where a bomb device may be activated

Excessive Traffic

Depending on an incident off site of some consequence, an inordinate amount of auto traffic may enter a facility's parking areas with people seeking shelter or assistance of some type. Occurrences such as this will cause bottleneck and back-up situations for responding emergency vehicles. Be aware that a security officer may be required to control the traffic in order to leave an open pathway for the movement of responding vehicles.

SECTION 6

Chapter 28

Familiarization with Criminal and Civil Litigation[67]

Introduction

Generally, loss-prevention, security, and safety directors are not familiar with criminal and civil litigation processes. Nevertheless, many of those who enter the private security and safety field originate from the public sector. These individuals are much more familiar with criminal litigation, but rarely have experience in civil liability and litigation. The following narrative will explain these processes, and why those professionals in loss prevention and safety must be aware not only of criminal prosecution, but how civil litigation could seriously affect a business or institution.

Because civil liability is such an important issue, the purpose of this chapter is to familiarize security and safety personnel with the legal significance and consequences of certain aspects of the work they conduct on a daily basis. The culpable actions by a security or a safety officer and the defensive remedies favorable to them are defined in civil law and are easily applicable to criminal law.

It is essential that all loss-prevention, security, and safety officers understand criminal liability and civil liability, and how such vulnerability—both criminally and financially—can affect them, their superiors, and the company that employs them.

The manager of risk, safety, or loss prevention must bear in mind that, along with the destruction of property, other serious consequences could significantly affect business and cause catastrophic financial loss. Along with a complete business disruption, the loss of good will, and the possibility of personnel without jobs, one of the more serious issues that could occur is the awarding of monetary damages in a civil lawsuit. Whether the award is compensatory and/or punitive, in many cases we will find that millions of dollars can be granted to a plaintiff.

Today, following any type of an incident, that person most probably will seek legal counsel in an effort to gain some satisfaction and financial award. An occurrence on or off premises that can be attributed to some fault of that company or its personnel may cause a person who has been aggrieved in some manner to attempt to seek redress in the form of damages. And in most cases, damages mean money.

In that regard, damages sought by the plaintiff in court can be negated or reduced when the company sued can show that its training, actions, or precautions were conditioned on its awareness and concern for the safety and security of the public and its employees.

Moreover, we must be mindful that a security officer is not a law enforcement officer, does not possess police powers, and has no more legal authority than that of an ordinary citizen. A security officer may be arrested and face criminal charges in a criminal court for any illegal action taken by the officer and not within the scope of his or her authority. Whether the security officer is exonerated in criminal court or not, he or she may still face civil litigation for any wrongs or perceived wrongs that a plaintiff believes were committed against him or her, and the burden of proof for a successful litigation is far less than in a criminal court. Acquittal on a criminal charge does not exclude the initiation of a civil suit.

Accordingly, we will acquaint the reader with an introduction into the varied processes of criminal and civil litigation so that you will be able to relate the following narration to your routine duties.

Awareness

Most civil lawsuits against security officers (and vicariously against their superiors) occur because of some overt act caused by a security officer that was outside of his or her ***scope of employment***. This may include an unwarranted assault or false arrest and detention.

The taking away of a person's liberty—placing that person in custody and detaining him or her—is a serious matter under any circumstance. Whether the arrest was by a police officer, security officer, or an ordinary citizen, certain elements must be present for any overt action to consummate a legal arrest. Whatever justification exists, the actions of the person who makes the arrest will be carefully examined later at length to determine whether such action was warranted and legal. Be assured that if the arrest was unfounded, litigation will be forthcoming against the person who made the "arrest," and an award in the form of punitive damages can seriously affect the officer and the business enterprise, both financially and in terms of good will.

The Application of Criminal or Civil Law

There are two types of law that control a security officer's actions and make him or her subject to criminal charges or civil litigation for a criminal, wrongful, or harmful act. Atonement or restitution granted to the plaintiff in a civil case is in the form of damages awarded, whereas in a criminal case, the offense is not against some victim (person) or entity, but against the state, and any penalty determined by the court will be paid by the defendant to the state.

To further clarify this point, civil law deals with the relationships between individuals, whereas criminal law deals with offenses against the state. In a civil action, the *plaintiff* is the person who has been wronged and who charges the person or entity (the *defendant*) for committing the "wrong," and seeks some satisfaction or atonement.

In a criminal action, the *defendant* is the person (or entity) who committed or had a part in the crime, and the *complainant* (the victim) is the person who has been offended or harmed in some way. The district attorney will act for the state on behalf of the complainant, and upon being found guilty, the defendant will have the sentence suspended, be fined, and/or be incarcerated. However, any type of restitution to the victim may not happen unless the court determines that it be made part of the sentence, such as the reimbursement of money stolen by the defendant. For a complete definition of the terms defendant, compliainant, and plaintiff, see Chapter 31.

An entity in a legal action may include a corporation, an organization, or any enterprise.

Chapter 29

Criminal Law

Criminal law (and liability) is concerned with the enforcement of legal duties to act, not moral duties to act. It includes an obligation to refrain from doing an illegal act or from failing to act as may be required. In other words, when one commits a voluntary act or an omission to act that is illegal as legislated in the law, that act is considered to be a criminal act. The basic element of criminal liability is conduct by an individual who, by his or her free will, either acts wrongly or fails to perform an act he or she is physically capable of performing. Both corporations and individuals may be found guilty of criminal liability.

In essence, criminal law includes those statutes dealing with crimes against the general public and individual members of the public, with penalties and all the procedures connected with charging, trying, sentencing, and imprisoning defendants convicted of crimes.[68] Accordingly, criminal law is statutory law, and the criminal justice system (police, prosecutors, judiciary, and corrections) will enforce the law and share a common goal to preserve the peace, prevent crime, and keep the community and its citizens safe. The person who commits an illegal act is arrested and brought before the criminal court to answer to the charge. Since the crime is an offense against the people of the state, any sentence or fine imposed on the offender is paid to the state.

In summary, criminal law defines those offenses against the state (the people) and prescribes the punishment for their commission or omission. Crimes are circumscribed by legislation (statutory law).

Criminal Liability

Criminal Act Defined

A crime may be defined as a violation of law by the *omission of a duty commanded* or the *commission of an act forbidden* by statute and conduct that unjustifiably and inexcusably causes or threatens substantial harm to an individual or the public interest. Criminal intent may not be required.

The loss-prevention and security manager, or the fire safety director, must be cognizant that any *act* or *omission* required or mandated by law by an employee or a company places that employee or company in jeopardy of criminal prosecution.

Liability and Culpability

A person is guilty of an offense when his *conduct* includes an *act* or *omission* and its accompanying *mental state*. Therefore, that person must have a culpable mental state where he or she intentionally, knowingly, recklessly, or with criminal negligence commits such an act or omission so dictated by statute. In essence, the subject must have intent to commit the crime, knows that it is a crime to do so, and has the ability and wherewithal to commit that crime. This is considered different from that person who was required to do or act as mandated by law or his position, and in failing to act as required (an omission), death, injury, or damage occurred.

Therefore, a person is culpable and subject to criminal and civil actions if he or she has one or more of the following mental states:

- *Intentionally*: By design, (the desire and the ability) the presence of will in the act that consummates a crime; has a conscious objective to cause a result or to engage in such conduct.
- *Knowingly*: With knowledge to commit or complete the act.
- *Recklessly*: Commits an act that he or she is consciously aware of and disregards a substantial and unjustifiable risk that such a result will or could exist; a gross deviation from the standard of conduct that a reasonable person would render, or a person that creates such a risk by reason of voluntary intoxication.
- *Criminal negligence*: Culpable carelessness; acting or omitting that which is required, and which a person of ordinary prudence would not have done or omitted to do.

Example

A person is hired as a trainman to switch tracks at various times and for various trains as required for the safe movement of these trains. However, because this person was inebriated, he or she began to doze and fell asleep at the controls. Subsequently, two trains on the same track collided; damage was severe, and deaths and injuries occurred. This person could be charged with the crime of manslaughter (among others), not because the occurrence was an accident or because there was an intent to cause harm on his or her part, but because the person did not fulfill his or her duties (an *omission*). Instead, the person acted in a negligent way that endangered the lives of others, causing death and injury, and that is the basis for being charged with a crime.

In the case of a *forbidden act*, if that same person—employed in the same position noted above, but with criminal intent—acted in such a way that two trains did in fact collide, causing death, injury, and damage to property, the criminal charge would be more serious. A person who had the knowledge, intent, and the wherewithal to commit this criminal act would be charged with the crime of murder.

Liability of Corporations

Any director, officer, or employee who is authorized to act in behalf of the corporation—or an officer of the corporation or any other agent in a position of comparable authority with respect to the formulation of corporate policy or the supervision in a managerial capacity of subordinate employees—is considered guilty of an offense when:

- The conduct consists of an omission to discharge a specific duty imposed on corporations by law.
- The conduct constituting the offense is engaged in, authorized, solicited, requested, commanded, or recklessly tolerated by the board of directors, officers, or managers acting within the scope of their employment, or on behalf of the corporation.
- The conduct constituting the offense is a misdemeanor or violation or a law defined by statute imposed as a criminal liability on a corporation.

Additionally, a person is criminally liable for any conduct that constitutes an offense that he or she performs or causes to be performed in the name of, or on behalf of, the corporation to the same extent as if the act or omission were performed in his or her own behalf.

Civil Litigation following Criminal Prosecution

Generally, a violation of criminal law does not exclude or otherwise affect any right or liability to damages, penalty, forfeiture, or other remedy authorized by law to be recovered or enforced in a civil action, regardless of whether the act or omission involved in such civil action constituted an offense in criminal law. Therefore, an acquittal, dismissal, or conviction of a criminal charge does not negate the possibility of a subsequent civil action by *either* party to the criminal action.

Chapter 30

Civil Law[69]

Civil law relates to private rights and remedies sought by a party who has suffered a wrong or has been damaged in some manner. When a person commits an act or fails to act as he or she may be required, when there is no right or privilege to do so, and when such act or failure to act injures the person, property, or reputation of another, either directly or indirectly, the act is a civil wrong. This is also called a tort, and the commission of a tort is a civil act and not a criminal act. Although a tort and a crime may be the same or similar in many cases, the parties, the burden of proof, and the damages are different.

Therefore, civil liability includes the procedures for civil law actions based on a violation of a legal duty or standard of care that may result from an accident, an omission of an act or duty, and an intentional or a criminal act that a plaintiff may have suffered.

Whether a defendant in a criminal action is found guilty or acquitted has no bearing on the ability of that defendant to initiate a subsequent civil action. Conversely, a security officer who faces a criminal charge in a criminal court, whether found guilty or not, may also face a civil lawsuit if the wronged party wishes to initiate such an action.

Most security personnel will be placed in some type of liability situation at some time in their careers, where a remedy will be sought by a litigant for a wrong or perceived wrong. Usually the litigant seeks damages in the form of money, but not always. A lawsuit may seek to change a procedure or a policy or simply correct a wrong.

In summary: Civil law is a vast body of statutory and common law with private rights and remedies available to a citizen. The basic element of *tort law* is a lawful obligation that is owed to the plaintiff (the wronged party) by the defendant (the party that caused the wrong).

The Question of Foreseeability

The courts have found that a corporation can be held liable for conditions on or within its premises, particularly where negligence can be established. Bear in mind that a person entering a premises open to the public has a reasonable expectation to believe that the premises is safe and secure. There is a duty for a corporation to randomly and routinely conduct inspections and examinations to find dangers that may befall the inhabitants. If and when these hazards or conditions are discovered, the corporation is obligated to correct or eliminate the danger in a timely manner. This will include alarm systems such as a fire alarm system, which must be reliable. If it malfunctions in any way, including failures in central station monitoring, it must be corrected in a reasonable amount of time. In certain facilities, as a further precaution, the local fire department may be advised that a fire alarm is inoperable.

A condition found to be a possible danger to employees, visitors, customers, patients, or students must be addressed as soon as possible so that the condition is corrected. To do otherwise places the corporation and the loss-prevention department at risk to criminal or civil liability.

Finally, as the loss-prevention manager, you must constantly monitor your facility, your employment and hiring practices, training practices, and any condition or risk that could harm any person who may wish to seek a civil remedy. This must be done so as to ensure that you and your department are doing everything legal and foreseeable to avoid or mitigate criminal and civil litigation.

Civil Liability

Negligence Defined

It is important at this point to define the term *negligence* as it relates to civil law. *Black's Law Dictionary* defines negligence as "the omission to do something which a reasonable man [sic], guided by those ordinary considerations which ordinarily regulate human affairs, would do, or the doing of something that a reasonable and prudent man [sic] would not do." This definition also includes the failure to use such care as a reasonably prudent and careful person would use under similar circumstances. It can be characterized chiefly by inadvertence, thoughtlessness, and inattention, i.e., an unintentional act by a person that does not come within the standards established by law for the protection of the public at large (*negligent wrong*). This is contrary to *wantonness* and *recklessness*, which are characterized by *willfulness* or *full awareness*, i.e., a voluntary or intentional act, either committed or omitted (*intentional wrong*).

Intentional and Negligent Wrongs Clarified

If a person suffers from some act, or a failure to act, when no privilege or right exists and when such act or failure injures the person, property, or reputation of another, directly or indirectly, the injury may be considered a *civil wrong*.

There are two types of wrongs: intentional wrongs (willful torts), and negligent wrongs (negligent torts). The plaintiff's complaint will note intentional and negligent wrongs, and it is possible that a security officer or other employee committed an intentional wrong, while the company may have committed a negligent wrong, and thereby lies the plaintiff's position for a claim of *vicarious liability*.

Consequently, a tort is a private or civil wrong for which the court will provide a remedy in the form of an action for damages. A personal tort involves an injury to the person or to his reputation and psyche, as distinguished from an injury or damage to real or personal property.

Negligence can be further defined in the sense that the business owner, facility manager, or the lessee of the premises failed to use *ordinary care* to reduce or eliminate an unreasonable risk of harm that was created by a condition on the premises that the owner, manager, or lessee knew or should have known about so as to exercise reasonable and ordinary care.

Ordinary care may be defined as the owner, manager, or lessee who *acted in a reasonable manner* in light of what he knew or should have known about a condition, and the risks that may be or have been caused by that condition. Moreover, the failure of the owner/lessee to use such ordinary care is *proximate cause* for the injury or damage, in that a natural and continuous sequence may produce an event, and without such cause, the event would not have occurred. Further, the plaintiff must show that for a *proximate cause* to be present, the act or omission that occurred must be such that an ordinary person would or should have foreseen that injury or damage might reasonably result from that event, and failed to use ordinary prudence or care under the circumstances.

The Importance of Documentation

Actions such as those described above demonstrate the importance of reports as well as documentation of routine audits and inspections within any institution or business establishment. Such documentation provides evidence of ordinary care in routine inspections and identifying and correcting security and safety problems.

Basis for a Lawsuit

In order to show a claim for security or safety negligence, the plaintiff must show that

- The company had a *legal duty* to take certain care and actions for the protection of employees, customers, visitors, etc., coming onto its premises.
- The company management knew, or should have known, that a crime or risk of injury or damage was *foreseeable* in the eyes of a reasonable person.
- The company failed to provide a *reasonable level of adequate security or safety* to prevent such a crime, risk of danger, or injury suffered by the plaintiff.
- The company's failure was a *proximate cause* for the damage suffered in the claim by the plaintiff.
- The plaintiff suffered actual damages and seeks redress because of the company's breach of its *legal duty.*
- The plaintiff suffered damages because of the actions (acts or omissions) of a security officer acting within or outside of the *scope of employment.*
- The plaintiff suffered damages because of the actions of an employee.
- A claim of vicarious liability can be made if the actions of the individual employee can also be attributed to other persons, including the corporation.

As far as the civil courts are concerned, the three elements regarding liability that will be examined by the court will be

1. Was there negligence?
2. Was there actual damage or loss?
3. Was the negligence the proximate cause of the damage?

Liability Contemplated

There are several ways in which a company, an institution, or a person may be held liable in a civil action. The following examples are based on actual incidents and are presented here to show the variety of issues and how they may affect an officer or a corporation:

1. The following are examples of intentional torts concerning the actions by a security officer:
   - ***False arrest, an illegal act***: A security officer arrested or placed a plaintiff into custody without cause.
   - ***False arrest and malicious prosecution***: A security officer failed to have probable cause for an arrest and detention, but continued with the process and falsely requested the police to charge a subject with a crime when, in fact, the security officer had committed a malicious act.

- ***Unlawful detention***: A security officer detained a defendant for an excessive amount of time, causing emotional trauma, prior to requesting the presence of the police.
- ***Violation of civil rights and defamatory remarks***: A security officer used derogatory or racial epithets against the plaintiff.
- ***Use of excessive force, excessive public exposure***: A security officer embarrassed the plaintiff in front of his family, friends, or passersby while consummating the arrest, thereby causing emotional trauma and a loss to the plaintiff's reputation.
- ***Assault and battery***: A security officer unnecessarily used excessive physical force upon the subject, causing an injury.
- ***Failure to act within the scope of employment***: A security officer, once proved to have knowledge of a safety violation or unsafe condition, failed to correct the condition or to act in a timely manner in safeguarding the plaintiff from injury.

Be aware that subjects who state that damages to their person, character, or integrity have been brought about by embarrassment, emotional trauma, a loss of personal reputation, or public humiliation can easily put forth a case in a civil court, whether or not the person charged with the offense has committed a wrongful act or has been found innocent or guilty of an offense in criminal court.

2. The following are examples of negligent torts:
   - ***A negligent tort***: During a report of a premises evacuation, which was a result of a fire in the building, the plaintiff attempted to exit the building via a fire exit visibly marked as such, but found it to be locked or blocked in some manner, thereby causing severe injuries to the plaintiff, who was unable to exit safely.
   - ***A negligent tort and a possible intentional tort by a salesperson***: An empty box was left in an aisle on the selling floor by a salesperson, and a customer tripped over the box and fell to the floor, suffering an injury.
   - ***A negligent tort (negligent hiring practices)***: A company can be held liable if it hires an employee without a thorough background check or overlooks certain prior behavior during the hiring process, and the employee thereafter commits a violent act while at work.
   - ***A negligent tort (negligent retention practices)***: A company can be held liable if it has information regarding violent actions or possible violent acts, threats by an employee, sexual or other type of harassment, and no action or very little action is taken by management.

- ***A negligent tort (premises liability):*** A company or institution can be held liable for damages in a civil court for any security issue where any foreseeable act can be established where the owner or lessee failed to use ordinary care to reduce or eliminate an unreasonable risk of harm, that was or may have been created by a condition on the premises which the owner or lessee knew or should have known about and corrected.
- ***Awarding of Damages:*** Upon the finding by a judge or jury for the plaintiff, the awarding of damages may take several forms. Usually the award is in the form of money, but not necessarily so. A plaintiff may seek retribution after having suffered a loss, detriment, or injury, or to have a procedure, process or practice changed or corrected. In addition, an award will fall under one or both of the following:
  - *Compensatory:* Those damages awarded as fair and reasonable compensation for pain, injury, humiliation, disability, expenses or some loss suffered by the plaintiff because of some action by the defendant.
  - *Punitive:* Awarded in reality to punish the defendant for wrongful actions and to serve notice as an example to others that such actions will not be tolerated.

Chapter 31

Possible Criminal and Civil Liability that Loss-Prevention Personnel May Encounter

Note that when civil litigation is initiated against a business owner or its agents (the employer's security/safety officer or other employee), the business is considered to be the *defendant* in the action. The person or group damaged is referred to as the plaintiff. The *plaintiff* is represented by an attorney, and the defendant is represented by the company's attorney or one provided by the insurance carrier.

This is in contrast to criminal law, where the person accused, arrested, and charged is the *defendant*, and the business, retailer, owner, administrator, agent, or security officer, as the case may be, is the *complainant* or the *victim*. The complainant or the victim swears before the court that an offense or crime has been committed and wishes that the defendant face the charge in court. The defense attorney represents the criminal defendant, and the prosecutor or district attorney represents the people of the state on behalf of the complainant or the victim.

The burden of proof in a civil case is a *preponderance of the evidence* (clear and convincing evidence—more for than against), and if so found for the plaintiff by a judge or jury, the plaintiff is entitled to money damages to compensate for the injury or offense. In a criminal action, it may be said that the *plaintiff* is the state (the people), represented by the prosecutor or district attorney (the state), and the person charged is the *defendant*. The injured party or the victim of the crime is the *complaining witness* or the *complainant*, and the state must prove its case *beyond a reasonable doubt*. The burden of proof is much more demanding than in a civil trial. If found guilty, the defendant may be sentenced to pay a fine or to a term of incarceration. This sentence is paid to the state, and the injured party or victim receives nothing, other than the knowledge that justice has been served. However, in some circumstances, the victim may receive some restitution or recompense from the defendant by order of the court, and in some states some type of victim compensation is offered.

If a person commits a crime that may also be a tort, he or she may be tried separately in either or both criminal and civil courts. The outcome of one court will not effect the outcome of the other court. Therefore, the complainant in a criminal action can also sue the other party civilly, whether found guilty or not of the criminal action. But note also that the criminal defendant, if acquitted of the charge, most assuredly will initiate a civil action against the complainant.

The implications and consequences of the actions or inactions of a security officer can be harmful both criminally and financially, not only to individual officers, but also to their employers or companies they may be contracted to serve. Without doubt, we can assume that most security personnel will be placed in some type of a liability or litigious situation at some time in their careers. The security officer should be aware that people don't care if a badge denotes "Security" or "Police." If they can, they will claim that their constitutional rights were violated or that some damage had been caused, and will attempt to seek some sort of remedy. The merits or quality of the case will not matter to an attorney, as most civil defendants settle before trial in an effort to free themselves from frivolous lawsuits that can tie up attorneys, witnesses, and businesses for long periods of time.

Chapter 32

Premises Liability

The Question of Security Negligence

A business that maintains a premises upon which another can be expected to enter is under the affirmative duty to make every reasonable effort to remedy conditions that could *foreseeably* create or contribute to a dangerous hazard or crime-inducing risk. Essentially, this has been interpreted as meaning that a *legal duty* is due those persons who enter upon a business owner's property by protecting those persons against harm and injury. If, as a result of this legal breach of duty, actual damages may have been suffered, the plaintiff may seek redress in a civil court.

Courts have awarded damages against business and property owners for injuries sustained under dangerous and threatening conditions, in addition to harm received at the hands of criminal assailants on the common-law theory of negligence. The relationship between the party who suffers an injury or some type of damage (the plaintiff) and the property/business owner (the defendant) will depend on the degree of care expected by the damaged party and the care required by the property/business owner.

The plaintiff, in order to initiate a civil action, had a legal right to enter and remain on or within the company's establishment. The establishment may be considered a *public place* open to the public in that the company may invite the general public at large to enter its premises for the purpose of selling its products or myriad other reasons for business activity. A theater, sports arena, or a museum, for instance, gives license and privilege to the public once a fee for entry is paid. After that fee is paid, unless notified of certain restrictions, the public will expect the same legal rights of security and safety. However, a person entering an establishment without paying the required fee commits a trespass.

Plaintiffs fall into three categories: *invitee, licensee,* or *trespasser.*

The Invitee

An invitee may be described as a customer, browser, or visitor who enters upon premises open to the general public. The public—whether a customer who is on the premises ready and willing to purchase merchandise, just browsing, a person who pays an entrance fee to enter, or a visitor who is present legally—is considered an *invitee,* and has *license and privilege* to be on the premises. A museum, a library, a supermarket, an office building with an open concourse or reception area, and of course a retail establishment would fall into this category of a place of public access. The invitee also includes the business visitor, a person who enters a business establishment for the purpose of some type of business dealings or a business appointment with the business owner or agent. An example might be a salesman who enters for the purpose of selling or offering a product or service. Accordingly, the business visitor also has license and privilege. Both the public and the business visitor are protected under the *invitee* concept.

In the case of a hospital or similar institution, limited access to certain areas may be prohibited or restricted to the public in order to control the security and safety of the facility and its inhabitants. However access is given, the greatest degree of protection is given to the *invitee* in regard to negligence that may cause some harm or injury.

The Licensee

This group includes contractors or workmen who are hired to come upon a premises for the purpose of repair, renovation, or other improvements. These workmen enter the business establishment with an implied or express consent, and are considered *licensees* but do not have the same protection as the *invitee.* A licensee has less protection, since any knowledge of a dangerous condition will prevent any recovery for damages if such would occur. The licensee is expected to accept the property as he finds it and look out for his own welfare. In addition, the knowledge of any danger that may be known to that licensee, who then enters the premises expecting the property as he finds and knows of it, may preclude the licensee from any type of recovery.

However, the business owner or agent has a legal duty to advise the licensee of any danger or dangerous condition that he or she is aware of prior to entering the premises.

The Trespasser

The trespasser enters a property unlawfully. This person does not have the implied or express consent of the business owner. He or she does not have *license* or *privilege* to enter thereon or therein. With just cause, a person may be given a trespass warning, orally or in writing, preventing that person from entering or remaining on or within a premises from that point on. Thereafter, if that same person is found on the premises at any time, he or she is considered present without license and privilege, and could be charged with a trespass. Accordingly, persons who fall into this category as trespassers and enter the premises unlawfully for the purpose of committing a crime therein, escalate the trespass offense to the crime of burglary.

Anyone who enters without consent, and without the right and privilege to do so, has no legal recourse if some injury or damage befalls him or her.

The owner or possessor of the property cannot be held liable for any harm or injury that may befall a trespasser. However, in rare instances a plaintiff, with intent to commit or while committing a crime on a business premises, suffered some injury, whether caused by his or her actions or not, and after a civil action subsequently received an award in court. There can be no reasonable explanation other than that a plaintiff's attorney has presented the more convincing argument or the reasoning that a jury may apply.

Anytime there is the question of an injury or damage that a victim can attribute to negligence on the part of the business owner, we can be sure that there is the possibility of civil litigation against that company in the near future.

Chapter 33

Vicarious Liability

Definition

Vicarious liability may be considered a form of *strict liability*, in that it has no element of fault. This is somewhat different from a liability such as *negligence*, that requires a *breach of duty*, and *intentional liability*, which requires *intent* or a *willful act*, and gives rise to an injury or a wrong. Simply defined, vicarious liability places the accusation of liability on one or more persons (or a corporation) for the actions of another.

If the plaintiff was harmed or wronged by an employee in the scope of his or her employment, redress may be sought against that employee *and* his or her superiors. Therefore, vicarious liability may be imposed on one person or company based on a civil wrong committed by another person. In any business enterprise, vicarious liability will most assuredly arise if the plaintiff can show that the business owner is or may be liable for the willful conduct of a security officer employed to protect the company's assets and acts within the scope of his or her employment. In some states, criminal laws may also apply in addition to civil liability, where the actions by an employee are held to be the responsibility of the business owner. An example would be a bartender who serves an underage patron who becomes inebriated and, after leaving the business premises, then kills or is killed in an auto accident with inebriation as a contibuting factor. Additionally, consider the case of a patron who consumes enough liquor to become inebriated with the knowledge or assistance of the bartender and who then leaves and causes a serious accident. Although this may or may not rise to the level of a criminal act, a civil action could be initiated against the bar owner and the bartender.

Similarly, if a security officer commits an act that may be considered to be wrong, the person who has been offended may seek redress in civil court for damages as a plaintiff. Therefore, if the security officer commits an offense or wrong outside the scope of his or her employment (intentional wrong), the company may wish to disassociate itself from the officer by terminating his or her employment and advising the employee to seek his or her own counsel. Specifically, if a manager, employee, or security officer commits a tortious act outside the scope of employment, and if the plaintiff cannot establish liability on the part of management or the corporation, that person is individually liable for the damages caused by his or her actions. This usually occurs in criminal actions such as a false arrest or an assault without cause. In this case, the company may not be obligated to offer legal counsel, and the individual alone may have to suffer any damage award.

If there is no *breach of duty* by the business establishment, but the security guard did act *within* the scope of employment by which there was an intentional or willful act and some harm was suffered, the plaintiff may sue the security officer as well as his or her manager, trainer, and employer. Further, the term *scope of employment* may be defined as any act or action that is intended to benefit or further the employer's business. Be aware also that acts forbidden by the employer or outside of the confines of company policy may be found to be within the scope of employment. As an example, the question may arise regarding the point at which the threat or use of force by a security officer in the performance of an arrest becomes excessive and thus departs from the protection of a company's assets. Such an argument may have to be determined by the court.

Essentially, the intent of a vicarious liability action is that damages awarded to the plaintiff against a security officer may not be as great as they would be if upper management or the corporation were included in the lawsuit.

Defense by a Company in a Lawsuit

The best defense against liability for any business establishment regarding safety and security issues, and its employees, particularly security or safety officers, is that ***due diligence*** be made in

- The routine investigation, inspection, and documentation of all safety and security hazards and incidents, with appropriate timely adjustments or corrections as may be required
- The hiring process
- The training process
- The subsequent effective supervision of that employee, with all the above documented.

See also "Record Keeping" in Chapter 36.

Chapter 34

Product Liability

Generally, product-liability lawsuits are directed toward any business that manufactures, assembles, wholesales, distributes, displays, sells, gives away, or in any way handles a product, article, or substance that causes harm to an individual or a group of persons. The product offered to the public for consumption must be considered free of defects, and if the item is found to be defective or to cause harm, the manufacturer must make good on that responsibility.

Other than the negligence noted under "Premises Liability" in Chapter 32, actions of this type seeking damages are not uncommon. One of the more litigious areas that a business owner may face is some loss or injury sustained by a customer that was caused by a product, article, or substance sold and carried by the merchant. This would include harm to a would-be customer while still on the retail premises caused by that product, article, or substance on display.

Definition

Product liability affects all parties along the chain from the manufacturer to the seller of any product that may have caused damage (injury) to another. Any product that is proven to contain inherent defects that causes harm to a consumer or user of the product, whether loaned, given, or sold, may be subject to a product-liability lawsuit. The claim may be based on negligence, strict liability, or breach of warranty. The injured party must prove that the product was in fact defective based upon a design defect, a manufacturing defect, or a marketing defect.

It is not uncommon today for a business enterprise to be held strictly liable for product-related injuries, even where the product had been purchased years ago by a plaintiff, was abused in some manner, used other than designed, or altered and made unsafe after the sale.

The plaintiff must prove a ***causal connection*** between the product's defect and the injury or harm suffered and that such defect did in fact exist when the product left the hands of the defendant.

The defendant will be held strictly liable for the ***unreasonable dangerous nature of the product***. Unreasonable dangerous nature may include usefulness and desirability of the product, availability of other and safer products that meet the same need, the likelihood of injury, and common knowledge and normal expectation of the danger, among others.

The loss-prevention officer should be aware of any product or merchandise on display that may come in contact with a customer or visitor on the premises, and how such contact could cause an injury. Such hazards must be removed or safeguarded so as not to place the business in liability. In a restaurant, a hazard may include a precarious or weak table or chair, or the unstable seating arrangements in a reception area that might cause an injury to the customer or visitor. The type of liability will depend on whether the hazard was unknown, known, or observed by company personnel and, if known or observed, whether action was taken to correct, remove, or make safe in a timely manner. Civil actions by a plaintiff regarding liability of a product once it is off the premises will initiate investigations by loss-prevention management and insurance carriers.

Liability regarding the product may occur under the following general areas:

- Based on the ***negligent conduct*** of the defendant in the manufacture or sale of the product.
- The ***implied warranty*** of the product, in that the manufacturer or retailer warrants that the goods are of quality and fit for the ordinary use intended.
- The liability of a product, which may be defined as the legal responsibility that a manufacturer, distributor, or retailer has to compensate persons who are injured as a result of using the product. This liability allows people to sue for damages when they have been injured or their property has been damaged because the product was ***defective***.
- Proof that there is a ***causal connection*** between the defect and the injury or harm suffered by the plaintiff, and that such defect did in fact exist when the product left the hands of the defendant.

Moreover, if the plaintiff suffers some misfortune based on a product, the plaintiff's attorney will attempt to affix liability regarding the product under the following general principles:

1. ***Design defect*** in that the product was defective or deficient in its manufacture or design, or was inappropriate for the purpose intended. Liability may also be based merely on the ***negligent conduct*** of the defendant in either the manufacture or sale of the product. In addition, the manufacturer, distributor, or retailer may be held strictly liable for the ***unreasonable dangerous nature of the product***.
2. ***Failure to warn*** in that the manufacturer or retailer failed to warn the consumer of known faults or defects that could harm the user, and failed to reclaim, recover, or remove from sale the defective product in a timely manner. This may include the continued marketing of a product that is known or should have been known to be unsafe.
3. ***Foreseeable misuse*** in that the manufacturer or retailer should have known that the product may be used other than intended or may be used more severely than the purpose designed. This may include the probability of counteracting the ***implied warranty*** of the product, where the manufacturer or retailer warrants that the goods or the product is of good quality and fit for the ordinary purpose for which the product is to be used.

[illegible]

Chapter 35

Contractual Liability

Rarely will the loss-prevention manager become directly involved with the insurance coverage that the business enterprise may require or the selection of the carrier who will underwrite the policy. That is usually left to the company administration. However, the manager should become familiar with the terminology of *contract liability*.

In essence, you and your company can be held liable for the negligent acts of another by means of a written or oral contract. In other words, liability for the negligent conduct of another person is incurred or implied through a contractual agreement. This will include all subordinates in your loss-prevention department and any contractual security officers you may hire from time to time. Many times these insurance contracts may automatically cover premises and operations coverage, but depending on the insurance carrier, some contracts may be more or less specific.

Contractual security guard companies will possess their own liability and worker's compensation insurance, and the business to which they are contracted should request and maintain an up-to-date copy of such insurance coverage. A business that contracts for the services of a security guard company should also obtain insurance coverage for any liable actions caused by that guard company, particularly if the actions can be vicariously proved. In addition, be aware that any workman given permission to construct, renovate, or repair on premises should provide necessary proof of liability and worker's compensation coverage.

Contract Law

Contract law governs contracts, which can take several forms. A contract can be formally written and described in detail; it can be written on a piece of scrap paper;

or it can be a verbal agreement. It is basically an agreement between two or more persons, describing an obligation to do or not to do something in return. To be legal, the parties to the contract must get something in return for giving something up for that return. For example, one cannot contract to give up a sum of money unless the other person gives up something in return. Moreover, the parties must be legally competent to complete the contract, and there must be no fraud or duress on anyone's part in the making of the contract. The agreement creates a legal relationship of rights and duties, and if broken, the law provides redress.

Because of the complexity of contract law, the loss-prevention manager should contact a corporate attorney for further information on this field of law.

Chapter 36

The Criminal and Civil Litigation Process[70]

Pre-Litigation

Preserving the Scene and Physical Evidence

In any serious crime, the local police jurisdiction, or in some cases federal law enforcement authorities, will become involved in the investigation and in preserving evidence at the crime scene. The forensic units of these agencies will have the expertise and necessary equipment to efficiently handle the task. In any catastrophic incident where death, injury, and damage are prevalent, these agencies will determine whether a crime was committed or if the occurrence was an accidental or natural disaster. They will have complete control of the scene of occurrence until they authorize the company to enter the scene and begin the recovery and restoration processes. The loss-prevention manager and his or her subordinates, along with company personnel, should assist and fully cooperate with these agencies.

Once all public safety authorities are finished with their duties of stabilization, recovery, and investigation, and company management has been given the authority to retake possession of the scene, certain actions by loss prevention will begin. The duties of a loss-prevention manager during and following a major emergency or disaster will be detailed in the Emergency Disaster Plan.

Security officers will be guided by police procedures or restrictions in the handling of any evidence or property that may be attributed to a criminal act.

Obtaining Admissions, Confessions, and Witness Statements

If an arrest is made and the subject in custody has, in fact, committed a crime, particularly where the security officer has observed the commission of that crime, or where a voluntary declaration of guilt is made, the admission should be reduced to writing.

If possible, a written and signed confession taken from the subject is the most desirable evidence of guilt. If a written statement cannot be taken, any oral admissions or statements made by the subject (or a witness for that matter) should be reduced to writing by the security officer as soon as practical. In any case, a witness should be present when the subject reads and signs his written confession or makes any oral admissions.

> A confession or statement should not be taken or received under threat or duress.

The *Miranda* warning need not be made part of the statement nor as a warning prior to any questioning, since security officers are not police officers. But if a police officer has knowledge of facts and circumstances possessed by the defendant and asks the security officer to take a written statement or an oral admission, the security officer is then acting as an agent of the police officer and must issue the *Miranda* warning. See "Color of State Law" in Chapter 37 regarding actions by a security officer when in the presence of or at the direction of a police officer.

Regarding witnesses, a written statement should be taken from any person having direct knowledge of pertinent facts concerning the crime or serious incident. If a written statement cannot be taken, the witness's identity should be noted for future reference, and any oral statements of importance should be attributed to that person and reduced to writing, and thereafter attached to the incident report.

If a security officer has the ability and is required to take a written confession from a defendant or written statements from witnesses, the officer should have some training as to the format and what is required to be included in the confession or statement so that the document will stand up in court.

Any statement taken from a pertinent witness, or in particular from a defendant, should contain the personal data of that person (i.e., name, date of birth, marital status, names of children [if any], spouse's name, home address, telephone number, how long at this residence, occupation and business address; gather as much identifying information as possible). At or near the end of the written narrative, a sentence should note that the statement was freely given, and that all that is contained in that statement is the truth. Moreover, witnesses to the reading and signing of a confession should indicate so by their signatures.

If the police become involved because the incident is criminal in nature, it is best to let the police investigate and not take confessions or witness statements. However, it should be noted that some police agencies look favorably on confessions taken by security officers prior to their involvement, as long as the security officer has the ability and expertise to do so. This is because a *Miranda* warning given to a suspect by the police might suppress any admissions that might be received in the interrogation process. Every statement becomes part of the case file and is subject to scrutiny by the defense attorney or the prosecutor at discovery or at trial.

Police procedures may vary with jurisdiction. It would be wise for the loss-prevention manager to determine what situations should be handled only by the police and what incidents can be handled by subordinate officers.

In any case where the crime is of a serious nature, the police will interview all witnesses, and the security officer will be of great assistance in identifying to the police all witnesses involved.

Contacts with or Comments to Unknown Parties

Any oral statements, however insignificant, made at the scene of an incident may be noted and used at a later time. The security officer should not express any thought, opinion, action, or response that will place the officer or employer in jeopardy of liability. Advise any associates assisting at an incident or at the scene of an occurrence to remain as noncommittal as possible while doing what is required of them.

Additionally, security officers must be aware that, following any occurrence or incident where an investigation or litigation may be forthcoming, they must be conscious of who interviews them or attempts to glean information through an apparent normal conversation. In essence, do not answer any questions from anyone unless you are sure that the person requesting information is an attorney or investigator representing the security officer, the employer, or the company that the officer is contractually assigned to. Naturally, while in court and under oath, the security officer will be required to answer all questions truthfully, unless upheld on objection by the attorney representing the officer or the company.

It is not uncommon for someone hired by an attorney for a civil plaintiff or criminal defendant to pass himself or herself off as an investigator or as a representative of your company's insurance carrier in an attempt to gather information helpful to the plaintiff's side. In like manner, during breaks or recesses in trial testimony, court observers or elderly retired people passing the time of day in a courtroom may engage you in a conversation you may wish never happened after you reenter the courtroom.

Record Keeping

Dependable record keeping by a loss-prevention department is a must. Security officers should complete a form with all pertinent known details for any incident that occurs. There are important reasons for the maintenance of records:

- ***Records retention as required by law or acceptable accounting principles***: Maintenance of accurate records over long periods of time, as required by law.

For loss prevention, this includes state, federal, and OSHA (Occupational Safety and Health Administration) requirements.

- ***Documentation of the incident or occurrence***: Reports of investigation, research, and inquiry for subsequent submission to the company's insurance carrier regarding possible criminal or civil liability or for production of such records in a court of law. This will include records and reports pertaining to customer or employee accidents, property damage, liability and civil actions, apprehensions and arrests, or any other incident that will or may require some future scrutiny.
- ***Inspection reports***: Documentation of routine and ongoing inspection of the facility for safety and security lapses or hazards, with appropriate timely corrections if needed. Records of this type are most important for production in any litigation process.
- ***Justification for the existence of a loss-prevention department***: Recording substantial and irrefutable evidence, compiled and statistically presented by the loss-prevention department, which can serve as justification for upper management to accept the need for a loss-prevention department. Formulating and presenting yearly, quarterly, or monthly analysis of all activities to administrative management will in effect establish a cost-benefit ratio. Justification could include:

  - Number of incidents or occurrences: by type, location
  - Customer, visitor, and employee accidents or illnesses: number, type, location, and disposition
  - Apprehensions and arrests: external/internal, type, property involved, disposition
  - Recovery and restitution: monetary/property/merchandise by value and type, external and internal
  - Civil actions: all litigation by type, pending or closed, and loss or recovery, if any.

Preparing Incident Reports

The security or safety officer must have the ability to record any incident that occurs during his or her watch, no matter how minor. For the more important occurrences, where times, dates, locations, notifications, people, facts, oral or written statements, etc., must be noted in detail for possible future reference, a timely incident report must be compiled. It should note who, what, when, where, why, and how, and the report should be complete as possible. The report should be maintained. If no number is assigned to the incident, such reports should at least be maintained chronologically. They should be signed and dated by the security officer completing the form. A master logbook detailing date, time, incident briefly described, and number assigned should be maintained for easy reference.

The security officer, the employer, and the company that the employer is contracted to, if any, will find that this report is most important if required for future reference, particularly in any criminal or civil court actions.

A security officer must be cognizant that the incident report or any written instrument must be as truthful to the facts as can be. The security officer must remember that he or she may have to swear before a court or hearing that everything contained in the report is the truth. In addition, conjecture or opinions should not be included in the report. The report should contain only the facts as reported or found by the officer. To do otherwise would place the experience, expertise, and professionalism of the security officer in question. Remember also that any report or writing the employer requires the officer to compile is considered a business record, and any untruth or fabrication contained therein may be considered a crime in many states. Additionally, along with the truth professed (sworn to) in the writing submitted, the charge of perjury may be considered.

If the security officer wishes to impart relevant information concerning the case that he or she believes to be important, but which cannot or should not be contained in the report, the officer should do so verbally to the appropriate people or office, usually prior to any court action.

Litigation

Criminal

The criminal process begins when a person is arrested and the complainant, the arresting officer, or the victim swears to an information (the complaint) before a judge requesting the suspect to appear and face the charge in court. It may also occur when a grand jury issues a *true bill* (an indictment) against a person who may or may not have been charged or arrested earlier. When the person or suspect has been arrested, that person is now considered the *defendant.*

Depending on the circumstances and/or the state in which the crime took place, one of the following may occur:

- A presentation by the grand jury with a *true bill* or an information may be sworn to prior to an arrest and the arrest made at a later time or place.
- The information may be sworn at the arraignment of the person arrested and, at that time, he or she will also be arraigned, with a trial date and bail set.

Civil

A civil case may begin in several ways. The attorney for the plaintiff may send a letter to the company advising that he or she represents a client (the plaintiff) who has suffered some damages and wishes to initiate a lawsuit, and that the company should contact its insurance carrier in an effort to provide information or to settle the case. This action may be called a letter of intent.

The attorney may also go directly to civil court to obtain a *summons and complaint*, in which the attorney describes the acts or offenses suffered by the plaintiff and the amount of damages (money) sought. The attorney, plaintiff, or a process server will serve the summons and complaint on the defendant—the person or company sued. The attorney may also obtain service via certified U.S. mail. In any event, once the attorney for the plaintiff makes known the intent to sue as described above, the process has begun, and the defendant, the company, or the insurance carrier representing the company must respond.

Discovery

Defined

Discovery takes place at a hearing, where the attorney representing either the defendant (criminal) or the plaintiff (civil) attempts to obtain as much information from the other side as possible. This will include all facts, witnesses names, statements, evidence, and anything else of importance that the adversarial attorney may need to build a strong defense or to present a more informed case.

In a criminal case, such hearings before trial may be for the purpose of determining whether the state has a prima facie case and whether admissions, confessions, and evidence were obtained legally.

In a civil case, the plaintiff's attorney wants to ascertain all the facts so that he or she can determine whether the proposed lawsuit is worthy, productive, and likely to be beneficial to the client in recovering damages. The plaintiff's attorney gains this information by depositions or examinations before trial.

Testifying at Depositions and Examinations before Trial

A deposition is formal testimony reduced to writing, signed and sworn to before a notary public. Basically, a confession does not differ from a deposition except that no notary public may be involved in a confession. Also, an employee may be required to submit to an interrogatory, which essentially is a set of questions in writing intended to be answered by a witness.

Additionally, an examination before trial (EBT) is a formal hearing, and may be considered a form of discovery. Basically, an EBT is testimony under oath, in response to questions asked by the opposing attorney with respect to the facts and circumstances surrounding the incident or occurrence. All sworn testimony is recorded by a stenographer. At this type of hearing all reports, photos, statements, confessions, or evidence must be produced or identified to the party requesting such disclosure.

Testifying at Trial

When appearing at any hearing or when testifying in court, the security officer should have a presentable appearance. He should wear a clean shirt, tie, suit, or sport jacket. The female officer should dress appropriately. The officer's shoes should be shined, and the officer should generally be neatly groomed. If the officer is required to appear in uniform, the uniform should be clean and neatly pressed. Jewelry such as lip, ear, or nose rings should be avoided. The officer should also arrive on time and be attentive and responsive. He or she should review all reports and statements to refresh his or her memory prior to testifying. Most probably, the officer will be required to consult with the attorney representing him and/or his employer concerning the case in question, his testimony, and what to expect from both sides. While on the witness stand, the officer should speak clearly and loudly enough to be understood by the judge, the attorneys, and the jury, if any.

Security officers must realize that their knowledge, demeanor, and presence in court will reflect upon any past actions they have taken. The more professional the officer appears, the more weight will be given to his or her testimony. Moreover, any formal reports compiled by the officer and submitted into evidence must be accurate, concise, complete, and easily read with as few mistakes in spelling and grammar as possible. A good report will also reflect upon the security officer's professionalism.

Chapter 37

Privacy Rights and Civil Rights Violations

Violation of Civil Rights

A civil right is an enforceable privilege that, if interfered with by another, gives rise to an action for injury. Examples of civil rights are freedoms of speech, press, and assembly; the right to vote; freedom from involuntary servitude; and the right to equality in public places.[71]

Discrimination occurs when the civil rights of individuals are denied or interfered with because of their membership in a particular group or class. Statutes have been enacted to prevent discrimination based on a person's race, sex, religion, age, previous condition of servitude, physical limitation, national origin, and in some instances sexual preference.[72]

A violation of a civil right is defined as any act that can be construed as denying a citizen of his or her civil rights as circumscribed in the United States Constitution. Statutory law elaborates upon those civil rights contained in the U.S. Constitution.

If evidence is presented that a private business has a policy of racial discrimination (presumed, unstated, or otherwise) against minorities such as Blacks or Hispanics, or any other type of discrimination (gender, sexual orientation), in hiring or in promotions, the business and those employees involved in such actions are subject to criminal and civil penalties.

Sexual Harassment

Sexual harassment is defined as any unwanted verbal or physical advance, sexually explicit derogatory statement or written material, or sexually discriminatory remark made by someone in the workplace that is offensive or objectionable to the recipient, causes the recipient discomfort or humiliation, or interferes with the recipient's job performance.

Title VII of the Civil Rights Act of 1964 describes the aforementioned behavior as sexual harassment and a form of sexual discrimination. The victim and/or harasser may be of either sex. The victim need not be of the opposite sex. The offender may be a supervisor, a coworker, or a nonemployee. The victim need not be the person offended; it may be anyone affected by the offensive conduct. In any event, the harasser's conduct must be unwelcome.

The loss-prevention officer must be cognizant that different people may perceive certain words, phrases, or actions (inappropriate or unprofessional behavior) as unacceptable and improper instances of sexual harassment. Therefore, the officer must be constantly aware the interactions between people, whether customers or fellow coworkers. The officer may also have to investigate a charge of sexual harassment by one coworker against another. In these instances, the officer will be guided by company policy and procedures, and may have to act in cooperation with the company's human resources department during the investigation and final outcome. In any event, the officer must be aware of the federal and state labor and criminal laws that define behavior that may be construed as an act of harassment.

The Civil Rights Act of 1964

In addition to the instances of sexual harassment noted above, the Civil Rights Act protects constitutional rights in public facilities (public education, public transportation, health care, public assistance, various social services, or any place of public access) and public accommodations (hotels, restaurants, and places of entertainment), and it prohibits discrimination in any federal program receiving financial assistance. It also prohibits unlawful employment practices for employers, labor organizations, etc., because of race, color, or national origin. It also creates the Equal Employment Opportunity Commission (EEOC) and its powers of enforcement.

The "Color of State Law"

Federal Civil Rights Act

Title 42, Chapter 21, USC Section 1983 (Civil Action for Deprivation of Rights)

Title 18, Chapter 13, USC Section 242 (Deprivation of Rights under Color of Law)

Other than the ordinary precautions for actions taken by security personnel, federal law (Title 42, Section 1983 and Title 18, Section 242 of the United States Code [USC]) may be of some interest to security officers who are privy to admissions or confessions under circumstances where such may be considered unlawful, or where such officers pretend to be or proffer themselves as police or law enforcement officers. Such behavior could cause a case to be lost, and the officer could face charges in criminal or civil court.

The basic provisions of Title 42, Section 1983 provide a right to sue state officials and others acting under the "*color of state law*" to deprive "any rights, privileges or immunities secured by the Constitution and Laws." Further, it provides that any person who, under the color of state law, subjects another to a deprivation of rights secured by the United States Constitution shall be liable to the injured party. In other words, the offender can be charged in federal court for the crime, whether or not he or she is charged in a state court, and whether or not found guilty in that state court. In addition, the offender may also be liable in civil court for damages. For example, a defendant charged in a state court for false arrest or assault can also be charged in federal court for violating the victim's civil rights. Double jeopardy does not apply.

To be in violation of this act, the violator must have acted under the color of state law. The statute prohibits those who act under the color of state law from violating a person's rights and privileges. It is intended to protect a citizen from illegal actions by a law enforcement officer or agency—the misuse of lawful authority. Therefore, the federal Civil Rights Act can rarely be satisfied in the case of anyone except a state or government official. But, if a security officer (or any person) acts *on behalf of, in cooperation with*, or *at the request of a law enforcement officer*, his or her actions are subject to the same restrictions as the law assigns to a law enforcement officer.

A Section 1983 action may be pursued against any defendant who *alleged to be a police officer* who used excessive force or a *citizen who collaborated with a police officer* in making an unlawful arrest. Additionally, the right to be free from false imprisonment is among the rights protected by the act.

The Americans with Disability Act (ADA)

42 USC Section 12201 et seq. Section 501 (1990) and Amendments

Employment (Title I)

The ADA prohibits employment discrimination by testing, by undue standards, or by job segregation, meaning that employers cannot inquire during preemployment whether an applicant has a disability, or the severity of such a disability, unless it concerns the ability to perform job-related functions. Under this law, a worker's compensation history is legal for use in the hiring process, and there are only three valid reasons to deny or rescind a job offer:

- The applicant's information on his preemployment interview is false
- The applicant is a threat to self or others
- The applicant is unable to perform essential job functions

Effective in 1992, the provisions of the ADA are applicable to any employer with 25 or more employees.

Public Accommodations (Title III)

All new construction and modifications must be accessible to individuals with disabilities. For existing facilities, barriers to services must be removed if readily achievable. Public accommodations include facilities such as restaurants, hotels, grocery and retail stores, and, in actuality, any building granting public access. Barriers within these facilities that prevent a disabled person from its use include restrooms, lavatories, stairways, elevators, doorway openings, and door handles, which must be corrected for any disabled person's use as long as it is readily achievable.

The right of privacy is an additional consideration regarding civil rights.

Chapter 38

Right To Privacy

Covert Surveillance

There will come a time where certain covert (a surreptitious surveillance enclosure) issues will arise, such as closed-circuit television (CCTV), two-way mirrors, coops, listening devices, or other types of surveillance or inspections that some people may consider an invasion of their privacy.

Basically, CCTV cameras in public areas, passageways, storage rooms, etc., can be used in the legal sense, in that employees and/or customers cannot expect the same right to privacy as they would have within a private dwelling or a public personal accommodation. Certain rights of privacy are given up upon entering a public building or someone else's property. An employee or customer may be considered a guest in the place of business, and the merchant or business owner may justify the use of cameras, two-way mirrors, or other types of surveillance based on the nature of work (for the employee); the necessity in combating and reducing theft of merchandise, property, and services; and the safety of employees and visitors.

However, there are certain restrictions that must be considered for the sake of propriety and decency when the use of cameras or one-way glass is contemplated. In the case of lavatories, restrooms, locker rooms, or dressing areas, male or female, customer or employee, an effective and valid case can be made concerning a right to privacy. No matter how serious a problem may be and the business owner's attempt to correct that problem, the possibility of some civil action or litigation because of an invasion of privacy, or the perceived notion of such invasion, is not worth the risk. Moreover, New York State's General Business Law notes that if a person enters a dressing room and is subject to being observed, the merchant or retailer *must post such notice outside of the dressing room.* In this way, the person may then have the choice to enter or refuse the use of the room. Of course, the retailer who has this policy may affect its business and good will in a negative manner. Therefore, it would be most prudent not to have any cameras or one-way mirrors in any dressing room for any purpose.

Check with your local and state laws regarding these arrangements that are considered an invasion of privacy and/or covert surveillance.

Wiretaps and Telephone Recordings

Eavesdropping Defined

Generally, most states have defined the use of a wiretap—the recording of conversations under clandestine circumstances—as *eavesdropping* when one unlawfully engages in wiretapping, mechanical overhearing of a conversation, or intercepting or accessing an electronic communication.

Moreover, the mere possession of any eavesdropping device, instrument, or equipment designed for, adapted to, or commonly used in wiretapping or mechanical overhearing of a conversation is a crime when the circumstances evince an intent to use or permit the same to be used in violation of law.

Wiretaps

Wiretaps on company telephones could violate federal and state laws, depending on how the procedure is set up and conducted. Although the employee's right to privacy may be balanced against the needs of the employer, consider this a no-win situation. If the loss-prevention department believes that a crime has been committed, and believes that one way to gain evidence would be through a wiretap, it is best to let the police handle it. Once the crime is reported to the police department, loss prevention could advise that a wiretap might be useful. The police will determine the benefits and legalities, and obtain the necessary court order to set up the tap.

> A security officer who sets up, condones, or is part of an illegal or improper tap or recording is subject to criminal arrest and civil litigation.

Many companies have a recording for incoming business calls that advises the caller that all conversations may be recorded for business or quality control purposes, and this is considered legal as long as the announcement is made before a conversation takes place.

This is not to say that a security officer cannot record a telephone conversation surreptitiously *as long as he or she is a party to that conversation*. Such a recording by a security officer where identification of the other person can be established, and where certain admissions are made by the other party, may be considered of great use as evidence at a later time. But be aware that there may be some restrictions in some states concerning the legalities of a recorded conversation that you are part of. Check with your state law, your company attorney, and company policy on any situation where any recording of a telephone conversation is to be conducted.

Employee Searches

The company may also have a policy concerning certain restrictions on employees, and as long as these policies are enforced equally, there is no loss of privacy. An example would be a rule that all employees must secure their pocketbooks, bags, valises, knapsacks, backpacks, etc., in their lockers and use a clear plastic bag to carry personal contents into the work area. Another example would be that all employees are subject to a bag check upon leaving the premises or work for the day. In this case, all employees leaving the work area must be examined; the security officer may not act in a random manner or subject one employee to the examination and not another. This cannot be considered an unreasonable search and seizure (U.S. Constitution, Fourth Amendment), since it is not an action by a government employee, such as a law enforcement officer, but by a private person, a security officer, as per the employer's written policy and procedures, which in effect were accepted by the employee as terms of employment.

Examination by a security officer of a private office or private desk, or opening an employee's mail, can be regarded as an invasion of privacy, since people have an expectation that these areas are private. Moreover, there is a greater risk of civil action if the security officer is working with or in cooperation with the police, since this will involve the issues of search and seizure, color of state law, and therefore, criminal liability.

Concerning employees' lockers, the employee must allow access if he or she has read and acknowledged written company policy that lockers are subject to inspection at any time. The employee may refuse such access, and no search of the locker can legally be conducted unless there is a serious public safety issue. However, the employee can then be subject to termination of employment.

Unreasonable search and seizures are protected by the Fourth Amendment and in *Wolf vs. Colorado* wherein the Supreme Court held that any evidence obtained in violation of the Fourth Amendment is inadmissible in a criminal prosecution.

In general, a security officer must:

1. Meet the test of reasonableness in determining probable cause
2. Prevent a general search using undue harshness

There are two types of searches that a security officer may conduct:

1. A search of the person that is incidental to a lawful arrest
2. A search performed with the consent of the person being searched, or consent to search property owned or under the control of the subject of the search

Chapter 39

The Invasion of Privacy and Defamation

Invasion of Privacy

There is no doubt that an investigation can cause emotional stress to both the guilty and the innocent. However, if a plaintiff initiates an emotional-distress lawsuit in response to an investigation, he or she must prove that the investigator engaged in some extreme or outrageous conduct that, in fact, caused severe emotional distress resulting in physical symptoms or bodily harm. Thus, every investigator should be cognizant of possible civil action alleging an invasion of privacy when investigating a person's pre-employment history, violation of company policy, or the report of a crime.

Publicity of private facts (invasion of privacy) can occur when an investigator makes a statement about an employee to another person not protected by ***privilege***. The employee may also believe that some intrusion takes place when an investigator looks into his or her private matters without having ***legitimate interest***. In other words, the investigator publicizes facts regarding another that are highly offensive to a reasonable person in which the investigator (a) does not have a legitimate interest or (b) passed on such information to one not protected by privilege.

The term ***privilege*** relates to those persons who have an intrinsic interest in or are privy to and have an interest in the investigation or its final results. This might include loss-prevention personnel, select administrators, or relevant managers.

The term ***legitimate interest*** can be defined as the total set of circumstances that would lead a reasonable, prudent, and professionally trained person to believe that an offense has occurred, is occurring, or will occur, and has the authority to investigate such offense. Basically, this could be less than probable cause but more than mere suspicion.

Procedures, expertise, and talent vary with different investigators, as does the type or seriousness of the investigation and how it is to be carried out. Therefore, the investigator should follow some simple rules in conducting a professional investigation:

- Although the security officer may initiate an inquiry or investigation into any process, procedure, possible or apparent loss, or suspicious activity or circumstances, the officer should have a ***legitimate reason*** to conduct an investigation of another person.
- The officer should respect the privacy and other rights of the people involved in the investigation.
- The officer should refrain from making any unnecessary comments or expressing opinions to anyone. This is not to say that security officers and their supervisors should not discuss among themselves feelings, thoughts, opinions, the direction or the target (suspect) in the investigation, evidence gained or to be gathered, or the complete process of the investigation.

Defamation

The issue of defamation must also be of concern to the investigator, since it may be the basis of a civil action or part of the litigation along with invasion of privacy. During any internal investigation, security officers may open themselves to an accusation of having defamed a person's reputation or character. When accusations or comments are made that are later proved to be wrong, and those accusations or comments are made outside of the ***closed circle*** (those protected by privilege) then ***the accused person—who, in fact, did not commit the act—has been defamed.***

Consider also that a defamation lawsuit can be based on two separate actions:

Libel: Where one produces or offers a malicious publication in print, writing, signs, or pictures tending to blacken the reputation of one who is dead or the reputation of one who is living

Slander: Defamation by words spoken; malicious and defamatory words damaging another

Libel covers the written word. Slander applies to the spoken word.

Chapter 40

Incident Investigation

Control of the Scene

Investigation is an important part of the job description of loss prevention and the duties assigned to that department. Whether the location of the occurrence is an accident or crime scene, certain precautions must take place, depending on the situation or the seriousness of the occurrence. Once the emergency situation is under control, such as a fire extinguished or an injured person attended to and removed from the scene, the security officer should then attempt to preserve the scene as much as it was when the occurrence took place. Since a scene of importance needs to be controlled as much as possible, the security officer cannot do all that needs to be done without some help. If other security officers are not present to assist, a responsible employee or manager can be utilized to help protect the scene.

The security officer should identify the complainant (if any) as well as all persons present during the occurrence, whether they witnessed anything of importance or not. If possible, attempt to separate witnesses to prevent distortion of facts. If necessary, and before taking any written statements that may be required, make sure the identification includes the name, address, age, date of birth, home and business telephone number, and a very brief recounting of what each witness observed and where he or she was when the incident took place. Such notations should be made in the officer's field notebook and maintained for future reference.

Anytime there is any remote possibility of subsequent civil litigation because of an accidental injury to a customer or visitor, or because of the severity of the occurrence or injury, photos of the scene should be taken. In the case of an injured employee, photos may be helpful to that employee or to the company, depending on the occurrence. And in the case of a crime scene, photos taken before police arrive may be of some importance, since a scene may inadvertently change or become contaminated because of people standing around and handling objects. If necessary, a diagram may be drawn of the scene, with measurements noting windows, doors, and important objects. The security officer should note everything of importance upon arrival at the scene and while he or she is there. The officer should describe the conditions upon arrival, and note whether certain items were out of place or appeared to be placed in position to accommodate a prepared scenario. The area of the scene should be searched thoroughly for any evidence, no matter how insignificant the officer may believe it to be.

Accident Investigation and Insurance Fraud

Cons, Scams, and Flimflams

Many of the small claims filed against a retail business may involve only a few thousand dollars; however, in total, they add up considerably. Although the merchant or business owner will have to absorb these petty losses because of the deductible on the insurance policy, the intervention of loss prevention and the insurance investigator many times can identify and deny claims made falsely by thieves.

Although slips and falls are usually staged at major retail, supermarket, and restaurant chains, they can occur at any location that can accommodate a large group of people. These claims of injury appear to be the most common type of insurance fraud, since it requires little time or effort to set the stage. The alleged "fall" or "accident" is most often without witnesses and difficult to disprove.

Additionally, there is often a delay in reporting the injury and claim. Because of this delay, the investigation is hindered by a lack of witnesses, the alleged accident scene has changed or has been renovated, and the subject has had time to set up doctor's examinations, hospital stays, and bills. The dollar amount of the claim is usually under $5,000, and many of these incidents are set up by a nefarious group of people known as "gypsies."[73] Because of the overwhelming volume of these types of claims by this group, the insurance industry finds that they are a tremendous problem to deal with. These gypsies are a mobile group and change names and addresses frequently as they move from town to town. The changing of names and other identification makes matching of these claims difficult for the insurance industry's Property Insurance Loss Register (PILR). The gypsy lifestyle and identification of their scams depends on secrecy. The gypsy claimant does not wish the scrutiny or publicity that a courtroom will bring. The other scam artist also does not wish publicity and usually will not involve an attorney. Therefore, both will want to settle rather than go to court. These types of cases are referred to in the insurance industry as nuisance cases.

Gypsies are many times implicated in stolen or bogus credit cards and bad-check scams, and flimflams at cash registers and returns counters.

The other scam artist is the so-called upstanding citizen who finds that, by staging a phony slip and fall, or by building upon a minor incident into a major injury, he or she can wait out the insurance company all the way to a civil trial. The claim is usually exorbitant, and the subject feels that, even if there is a settlement just before trial, he or she will come out ahead. This subject most often has representation by an attorney and the services of a doctor, both of whom may be as shady as the subject. Again, many times the report of the alleged injury is made at a later time after the alleged date of occurrence.

Caution should be taken, however, concerning the honest customer or visitor who has in fact sustained a bona fide injury. In aiding an injured person and the subsequent accident investigation, all cases should be handled with equal consideration and appropriate professional conduct. Whatever you may believe concerning the facts, respond as though the incident is a bona fide accident. Make no personal comments that might be litigious at a later time. Also, this admonition is to be applied to any coworker present or assisting at the scene, because one never knows who may be listening.

Early detection and thorough investigation of claims having involvement by certain groups such as gypsies have proven to be the best defenses in defeating fraudulent and non-meritorious claims. Once an accident and injury has been reported, loss prevention should examine the scene immediately for witnesses, inconsistencies in the injured person's statements, and physical evidence observed at the scene. Photographs and measurements should be taken of the accident scene. Comments by the subject, however minimal, vague, or inconsistent, should be noted as soon as practical. Complete identification should be gathered from the subject, particularly from photo ID of an official nature such as an operator's license. Include the social security number, business address, and, if possible, the previous home address. Note whether any of the corroborating and "overly cooperative" witnesses present are friends or relatives of the subject, and properly identify them and all witnesses for future reference. If the security officer has any suspicions of a possible fraud, he or she should advise the insurance representative or assigned investigator upon submission of the accident report.

Conclusion

Make sure that all required information is obtained so that an accident report may be completed soon after the incident. Many times the injured party or a relative at the scene of the accident, or a legal representative at a later time, requests a copy of the official accident report. Respectfully deny that request. You could offer pertinent information, names of witnesses, and the name of your company's insurance carrier, but no inhouse or insurance company accident form should ever be supplied to anyone. State that only the business owner's insurance company representative or attorney representing the business will have that right of disclosure. The company and its agent have the legal right to deny a request or demand of this nature. In fact, the only time a copy of an accident report will be offered will be at the request of the plaintiff's attorney at an examination before trial or at a discovery hearing.

Epilogue

The Weakest Link in U.S. Security

On September 11, 2001, an attack on the United States found that private security services were involuntarily transformed from an army of "rent-a-cops" to protectors of the homeland—or so we believed. Immediately after this attack on the security and safety of our homeland, protection became the linchpin to security awareness and the exceptional growth in the recruitment and employment of security guards. At least that was the enthusiasm for the first two years after 9/11. Since then private security has returned to the mindset of pre-9/11. American business has fallen back in preparation, employment, and held salaries to levels prior to the World Trade Center tragedy.

Federal and local law enforcement are to be commended with recent arrests here and abroad of terrorists, both foreign and domestic, before they were able to commit any act of aggression. However, even with dire warnings that the 9/11 event was a harbinger to that which is yet to come, American business still lacks any commitment to ensure the security and safety of their companies, their employees, or the general public.

The Issue of Wages and Aptitude

"The effectiveness of critical infrastructure guards in countering a terrorist attack depends on the number of guards on duty, their qualifications, pay and training."[1] Critical infrastructure is defined as "systems and assets so vital to the U.S. that any destruction or incapacitation would have a serious impact on the security, economic stability, health and safety of our citizens."[2] Moreover, "Overall employment of U.S. security guards has declined in the last five years. Contract guard salaries averaged $19,400 per year in 2003, less than half of the average salary for police and well below the average U.S. salary for all occupations."[3]

In order to professionalize the security vocation, some states have enacted minimum standards of training and background. In New York State, for instance, security guards must be certified by the state prior to any employment in security, with a complete background check, including a fingerprint check at the federal and state level, and the required minimum training. Training consists of 8 hours preemployment preparation, at which time a person may apply for a security position. Shortly after employment, the guard must complete 16 hours of training in law, management, demeanor, etc. Thereafter, in order to maintain this state certification, the guard must take an 8-hour in-service course yearly. Armed security guards are required to also attend these courses, maintain a "carry" pistol permit, plus attendance at a 40-hour pistol training course. The 8-hour in-service course and the 40 hours of pistol training are mandated annually. The cost for this training, certification, and licensure is borne by the applicant. With all that is required for a candidate to apply for a job in the security field, one wonders about the quality of the personnel employed. A fine is prescribed for those security companies and guards who fail to abide by the law and never apply for or renew their licenses as required. In order to fill positions, particularly at a moment's notice, many security agencies will hire officers without licenses, believing that, if they are found to be in violation, any fine imposed will be minimal. Unfortunately, this law is rarely enforced unless some criminal action takes place on the premises of the guards' employment or a guard becomes involved in some type of criminal activity.

Accordingly, recent studies have shown that, nationally, the average training for security personnel is exceedingly below par, and that huge turnover rates for this group have only added to the low quality of service provided.[4]

Immediately after 9/11, New York City underwent a brief, colossal movement toward improving its security standards. Unfortunately, the number of security guards soon fell back to the pre-9/11 level. A lackadaisical attitude began to develop in America. In a shocking revelation, the public advocate for the city of New York found that, in a city survey concerning a cross sample of security guards from the five largest providers of security services, the following statistics were obtained:

- 12% of security guards surveyed reported having no training.
- 17% had less than the required 8 hours of pretraining.[5]

New York State requires that a security guard obtain 8 hours of training prior to hiring and an additional 16 hours of on-the-job training shortly after employment. However, the enforcement of provisions concerning penalties of the various sections of the New York State General Business Law (Article 7A; Security Guard Act) has been less than desirable.

This problem is not confined to East Coast cities like New York. The economy has affected training standards, with little improvement since 9/11 or the aftermath of Katrina:

- 46% of California security guards surveyed noted that they received no training from their employers prior to hiring.
- 52% said they received no training in emergency response.
- Turnover rates for security personnel in Los Angeles ranged from 93% to 243%.[6]

Moreover, according to a nationwide survey by the Associated Press, the security industry is governed by a maze of conflicting state rules and regulations as to minimal requirements, training, and background checks. Tens of thousands of guard applicants were found to have criminal backgrounds.[7] While major cities have a ratio of three or four security officers to each police officer, minimum wage is the norm for a security officer. Working 40 hours a week and often being forced to work overtime hours, a weekly wage of $300 to $350 is not uncommon, thereby recruiting only the most uneducated, untrainable, and/or unreliable. Cutthroat competition by security firms attempting to win contracts with low bids do so because business is looking for the cheapest available, which in turn has kept the wages low, with applicants below par and high-level training nonexistent.[8]

As a certified New York State security guard instructor until recently, the author found that these statistics are not inaccurate. Many applicants for New York licensure who completed the initial 8 hours of training never returned to any authorized security training school for the additional required certification. Again, enforcement of the New York Security Guard Act has been found to be insufficient at this time in our history of terrorist activity.

The Case for Contractual or Proprietary Security Officers

The probability exists that, whether the security force employed by a business enterprise or institution is contractual or proprietary (in-house), without the necessary incentives, the security personnel will be poorly paid and poorly equipped and subsequently will provide a low level of service.

Many security or loss-prevention managers have known for years that, in hiring an outside security agency "you only get what you pay for." Many security agencies have developed efficiencies so as to compete in a low-margin, volume-based business in order to gain contracts and still make a modest profit. These efficiencies, unfortunately, have degraded the security service by cutting corners in order to compete. Reducing office overhead, supervisory personnel, acceptable uniform appearance, and liability insurance are a few reductions that are not uncommon. Moreover, lowering wages or keeping them status quo, thereby continually having to hire low-level applicants reduces the quality of the officer.

Businesses tend to look for the best service but also the cheapest. Security agencies become involved in bidding wars with their competitors in an attempt to gain contracts. A business that is willing to pay for the lowest level of security that is available will get what it pays for. Assigned posts will not be manned because of tardiness or absenteeism, supervision becomes inferior or nonexistent, sloppy or poor uniform practices degrade the image of the business, and civil and criminal liability issues arise based on poor training or performance.

On the other hand, a business that has complete control (proprietary) of the hiring practices, level of wages, quality and aptitude of personnel, and training and promotional pursuits can be expected to have a security force that will offer the quality of service demanded. The difference between the services offered by a security officer can be remarkable in appearance, professional responsibilities, and how their tasks and functions are carried out. Consider, however, that compensation will also have an effect on the quality of any service.

Whatever the type of service expected, costs will not only be limited to wages and training. Take into account assorted government taxation; varied insurance coverages; uniforms; hiring and background services; acceptable supervision; and acceptable level of officers to cover vacations, days off, sickness, and training, along with medical, vacation, and severance pay, if offered and available.

Many security agencies are well-prepared to offer quality service at a fair price, many times equal to in-house services, and refuse to reduce their presentation by becoming involved in a bidding war. On the other end, all that is required is that commercial business establishments offer compensation equal to the services they require. The difference between the services offered can be dramatic and notable if a business or institution is willing to pay for the service that is expected.

Governmental Services

Critical-infrastructure security guards in the federal service require training only in the airport screener and nuclear employment sectors. Twenty-two states require basic training for licensed security guards, and only a few offer any type of counterterrorism training that is minimal at most. Twenty-eight states require no training prior to or after hire. In 2003, there were approximately one million security guards (including airport screeners) employed in the United States, compared with 650,000 police officers. Nearly half of these guards were employed directly by the institutions they served; the rest, approximately 53%, worked for contract companies providing outsourced guard services.[9]

Federal, state, and local law enforcement, along with local fire services, are our first lines of defense against any disaster or catastrophe. However, during a severe emergency and depending on its extent, public authorities can be expected to be overcome quickly. Recently, public authorities have recognized that private security and loss-prevention officers can offer their assistance and expertise, particularly during the preliminary response to an incident, until public safety professionals can manage and take control. Nevertheless, emergency service of this type offered by the private sector must depend on their qualifications, training, and experience. Without decent compensation, private security will never reach that level of expectation.

Until the demand for better security is required by a business community that is ready to make a commitment to that end, and the public is willing to accept the price for this rise in quality and service, security and safety protection will remain stagnant. It is hoped that this book adds to the enlightenment, professionalism, and credibility of the risk management, loss prevention, security, and safety industry and serves as an addendum to the body of knowledge on emergency and disaster management.

Notes

1. "Guarding America: Security Guards and U.S. Critical Infrastructure Protection," CRS Report for Congress, November 12, 2004; Congressional Research Service, The Library of Congress, prepared by Paul W. Parfomak, Science and Industry Division.
2. Ibid.
3. Ibid.
4. "Disturbing Trends…," *Loss Prevention Magazine*, pp. 97–98, May/June 2008 issue, by Security Resources—Security and Investigative Solutions, Cherry Hill, NJ.
5. Ibid.
6. Ibid.
7. "Low-wage security has high cost—insecurity. Guards with little training may be weak link in homeland security," *The Orlando Sentinel Newspaper*, March 30, 2007, by Larry Margasak, The Associated Press.
8. Ibid.
9. CRS, "Guarding America."

The only defenses against terrorism in the *short run* are interdiction and self-defense.

The best defense against terrorism in the *long run* is building up secure civic infrastructures.

Just War Against Terror—The Burden of American Power in a Violent World, by Jean Bethke Elshtain,

©2003, p. 154, published by Basic Books, New York, New York, 10016.

Appendix A

Emergency Procedures Summarized

The following examples of procedures, and policies may be used as a model for the development of a policy or procedure statement.

The procedures to be followed in the various disasters or emergencies that may occur in or around a business establishment should be made part of the company's Emergency Procedure Plan. Moreover, these procedures should be part of the initial and ongoing training for all personnel, particularly those assigned to the loss-prevention department.

General Evacuation Procedures

When the alarm for an evacuation is initiated, all company personnel should comply with the following procedure:

- Shut down all hazardous operations.
- Remain calm and leave the area in an orderly fashion.
- Follow instructions and established evacuation routes.
- If necessary, assist all disabled persons during the evacuation.
- Close all doors as you leave, but do not lock them.
- One person should be assigned the responsibility to be the last to leave his or her department, section, or area in order to determine that all employees, visitors, customers, or contractors have left the area.
- Upon leaving the facility, move away from the building and assemble at a predetermined area assigned to you or your department for a head count.
- Do not block the streets or driveways. Emergency vehicles must have free clearance.
- Stay at the assembly area until directed otherwise.

Fire Safety

If the company has compiled and advocated a fire safety plan, all employees should be instructed in the procedures and responsibilities contained in this plan. However, if no written plan exists, and in the case of a reported fire, the employees should adhere to the following procedure:

- Report or sound the alarm.
- Notify the fire department.
- If the fire is small, try to extinguish it with the proper type of extinguisher.
- Do not jeopardize personal safety.
- Attempt to disconnect energized electrical equipment if safe to do so.
- Do not allow the fire to come between you and an exit.
- Do not break any windows, do not open any doors that are hot to the touch, and close all doors behind you as you leave an area.
- Do not use elevators.
- If the fire cannot be extinguished prior to the fire department's arrival, follow the general evacuation procedure.
- Leave immediately when the evacuation announcement is made. Do not attempt to save personal possessions.

Hurricane, Severe Storms, and Floods

Generally, the following items should be considered:

- Keep up to date with local radio and TV weather reports and/or NOAA Weather Radio for instructions.
- Plan ahead before the storm arrives.
- Tie down all loose items located outside of the building or move them indoors.
- Check battery-powered equipment and backup power sources.
- If the storm is severe, disconnect all electrical equipment and appliances not required for emergency use.
- Do not use the telephone except for an emergency.
- Avoid structures with wide roof spans, such as warehouses, auditoriums, etc.
- If personnel are to remain inside, locate them within interior rooms and spaces.

Hurricane

- Board up windows or protect them with storm shutters or tape.
- Open windows slightly on the side away from the direction of the storm to equalize the air pressure.
- Leave low-lying areas that may be inundated by high tides or waves.

- Stay in the building if it is sturdy and on high ground unless local public safety authorities order an evacuation to designated areas or shelters.
- Stay indoors in interior spaces and away from windows.
- Do not be fooled by a lull in the storm, e.g., the calmness of the hurricane's eye. The other side of the storm may be even more severe.

Excessive Rain, Snow, or Flooding

- Move to upper floors if waters begin to rise.
- If a flash-flood warning is given, evacuate the immediate vicinity.
- Follow instructions of public safety personnel.
- During flooding or excessive rain, know the depth of water or dips in the roadway before driving through.
- Do not enter water above your knees.
- If your vehicle stalls, abandon it and seek the high ground.
- Do not reenter an area previously deemed unsafe until directed by public safety personnel.
- Be aware that excessive rain can cause a saturation of the ground in certain areas and could cause landslides.

Earthquake and Tornado

These emergencies occur with very little warning, if any.

Earthquake

During the shaking:

Indoors

- Stay where you are during the occurrence.
- Take cover in a supported doorway or under sturdy desks, tables, or any furniture that can protect you from falling objects.
- Stay away from glass windows, doors, display cases, etc.
- If possible, stay near the center of the building.
- Do not run to stairways or fire exits. They may be damaged or jammed with people.
- Do not use elevators.
- Do not use candles, matches, or other open flame; there may be gas leaks.
- Extinguish all fires with the proper extinguisher.
- Once the quaking ceases, attempt to leave the building with caution and assist others who may have been injured.

Outdoors

- Move to an open area away from buildings, utility poles and wires, trees, etc.
- If there are no open areas and you are forced to stand near a building, watch for falling debris. Most injuries are caused by fires and building collapses.

If driving a vehicle

- Stop the vehicle as soon as safety permits.
- Avoid overpasses, underpasses, bridges, power lines, utility poles.
- If possible, park vehicle as far as possible from tall buildings.
- Remain in the vehicle until the quake ceases.

Tornado

- When a tornado is imminent, immediately locate all personnel in the basement if available, and if not, seek an interior hallway, a closet, or room with strong walls.
- Upper floors are unsafe.
- If caught out of doors and time permits, seek safety in a ditch or ravine, under a bridge or overpass, or under a vehicle. Do not remain in a vehicle, but as a last resort, crawl under the vehicle.

Volcanic Eruption

Although rare in the United States, those residents living and working in the vicinity of an active volcano must be aware of certain characteristic activity distinctive to an eruption and the reactions that will be expected to occur by the general public during or following a volcanic blast. Other than a lava flow in the immediate vicinity of the volcano and the rush of superheated clouds, falling debris and ash may cover large areas, causing death and injury unless some protection is offered. In general, an evacuation of the immediate area would be prudent unless directed otherwise by public safety officials.

Hazardous Material Accident

During a hazardous material accident or spill:

- Evacuate the immediate area.
- Initiate appropriate first-aid procedures for people that may be injured.
- Secure the area as much as possible.
- Notify appropriate support services: fire department, utility gas company, police and fire emergency hazmat units.
- Do not reenter the affected area until authorized by public safety personnel.

- Do not attempt to assist in the clean up of a hazardous spill or substance unless you are trained in such operations and have the necessary protective equipment.

If the hazardous-material accident occurs in the local community or in close proximity to the business establishment:

- If possible, listen to local radio and TV announcements.
- Follow instructions of all public safety officials.
- Evacuate when directed; follow designated evacuation routes to safe areas.
- Do not reenter the affected area until authorized to do so.

Utility Failure: Disruption or Severance

Blackouts and Brownouts

Because of the excessive use of electric power and the failure of power companies to keep up with the demand, brownouts and blackouts are no longer uncommon. Generally, these incidents will occur unexpectedly. When they do:

- Remain calm.
- Remain where you are and open all available blinds, shades, and curtains so as to receive more outside light.
- If you are in an unlighted area, attempt to go to an area equipped with emergency lights.
- If the telephones are working, report the outage to the proper utility.
- Do not use candles or an open flame. Fires can easily be started under these circumstances.
- If directed to evacuate, use caution, and assist all disoriented and disabled persons.
- If trapped in an inoperative elevator, stay calm and use the emergency intercom or bell to alert security or emergency personnel of your situation.

Gas Disconnect

A disruption in gas service can be a major or minor obstacle for the facility. Causation of such a loss can range from a break in off-premises gas lines due to some type of an accident to an outright sabotage or terrorist act. If the loss is due to an accident, the local gas utility would surely be advised immediately by various sources, since a gas leak presents danger of fire or explosion.

If the cause is human-made, we can assume that the damage to gas lines would be substantial and that the utility would need some time for restoration. In that event, a major evacuation of the area by civil authorities will take place until it is safe to return.

Water Loss

Water loss would cause some inconvenience until the utility made the necessary repairs. More significant would be an action by a terrorist in which a biological or chemical agent is infiltrated into the water supply, most probably at the source or treatment stations. Unfortunately, because such an act would be surreptitious, mass infections, serious illnesses, or death would affect the populace before the source could be identified. The civil authorities will control the occurrence, and all loss-prevention personnel will be guided by their direction.

The Terrorist Threat

We must be constantly aware that any type of attack may be terrorist related. If so, we can expect that such a cessation of one or all utility services will be long term, causing a severe hardship on the populace.

Follow the emergency procedure plan.

The Basics of Disaster Management in the Business Community

A disaster is an unplanned occurrence that disrupts the routine operation of a company; damage, injury, or death may occur. It may include the following occurrences:

Internal: Fire, chemical/toxic spills or exposure, gas leaks and explosions, structural collapse or damage, equipment failure, larceny, fraud, etc.
External: Bomb threats and explosion, all types of terrorist and other criminal activity, major assaults, and natural disasters

Because a business exists to make money, a disaster can cause a major loss of equipment, stock, productivity, and service that will affect the bottom line. The following represent areas of concern for the loss-prevention or safety professional:

The Foundation of Disaster Response

In order to identify and respond to an emergency or disaster, we must:

- Understand what constitutes a disaster and how it will affect the company.

- Know how to respond to disaster situations.
- Know who is to be involved in the disaster response.
- Prepare and train for disaster response.
- Request appropriate assistance from emergency and/or public agencies.
- Treat the injured and secure the scene as best as possible.
- Implement the disaster recovery process.

Preparation and Planning

- Assign the person who will take charge of the command post.
- Identify the key management personnel for the command post.
- Identify who will coordinate that position until arrival of the post supervisor.
- Recognize that the supervisory security officer or director of the command post will be the person to request outside public safety assistance.
- Determine the key areas of protection; prioritize the areas of concern.
- Identify the key personnel who will respond to various emergencies or occurrences, and their responsibilities.
- Recognize that, initially at least, backup assistance will be minimal.
- Recognize that security officers will be the initial emergency responders for first aid, fire response, etc.
- Care for the injured.
- Locate and use applicable emergency equipment.
- Maintain control of the premises.
- Identify the responders: who is to be involved, and in what type of occurrence.
- Identify the disaster team leader who will be in charge at the scene.
- Identify the key support personnel: maintenance personnel, housekeepers, employees trained as EMTs, or first-aid response.
- Identify the media spokesperson.
- Coordinate and cooperate with local public safety authorities: police, fire, ambulance, hazmat units or services, Office of Emergency Preparedness. Include also utility companies; electric, gas, telephone/telecommunications, and private emergency support, alarm and central station services.
- Identify transportation services.
- Identify radio companies and services.
- Identify glass replacement, structural containment, and contractual recovery and restoration teams.

Training

The ability to identify and respond to a disaster or emergency is one of the more important roles for a security or safety officer. Training, experience, and company policy as well as local, state, and federal laws should be taken into consideration where applicable.

Internal Activities

- Effective use of and constant review of the Emergency Procedure Plan
- Assignment of responsibilities
- Location and use of all emergency equipment
- Normal and alternative communications systems
- Key areas that require essential protection
- Conflict resolution strategies
- Constant and documented drills for all types of emergencies

External Activities

- Assorted training from outside public safety authorities:
  - Police department: bomb threats, response issues, etc.
  - Fire department: types and means of response, fire inspections, etc.
- Awareness of the fire response plan set by the company in coordination with the fire department and central station
- Awareness of hazmat response and initial duties at the scene of a chemical spill or toxic exposure
- Awareness of personnel response to civil disturbances and criminal activity

Recovery

Unless management determines otherwise, all areas must be capable of normal operation. If unable to initiate a normal mode of operation, partial operation of the business may be the only appropriate or acceptable action.

- All personnel must be advised of the "all clear" at the same time.
- Security and safety personnel should participate and be part of the recovery team.
- Assess the damage.
- Recover as much property as can be salvaged.
- Document and photograph all property that will be written off as a loss or covered by insurance.
- Loss-prevention department shall document in writing the occurrence and all activity surrounding the incident. (Prepare a detailed incident report.)

The Disaster Management Process Summarized

Other than the training process for first aid and fire brigade responders, and depending on how catastrophic the occurrence may be, the loss-prevention manager may wish to apply the following procedures.

Priority

The company, its administrators, and its managers must be committed to safety and security, with the assignment of designated employees to act in the various events that may occur.

- The business entity must be committed to respond to any emergency or disaster.
- The response is made for the protection of life and property.
- Disasters and emergencies, and the response to each, must be defined in terms relevant to the particular business enterprise.
- An Emergency Procedure Plan must be instituted to include all phases of disaster and emergency response within the business establishment.
- Disaster management and response must be made part of that Emergency Procedure Plan as company policy and procedure.
- The business entity must provide the necessary training of personnel and equipment required for such a response.

Responsibility and Planning

The person most relevant in the management and control of this area would be the loss-prevention manager. Therefore, the following details would be of great concern:

- Learn as much as possible regarding terrorist and violent activity that could take place at or within the place of business, and train for such an occurrence.
- Training for various disasters and emergencies is an important function of management.
- Identify those areas of vulnerability in and around the business establishment.
- Identify present security and safety measures and procedures, and correct or install those that may be required.
- Review, train, and review again all emergency evacuation procedures.
- Provide written policy and procedures for handling the various incidents that may take place.
- Provide detailed assignment and responsibility of crisis team personnel:
  - Identification and structure of a command center or post, and the proper equipment for operation
  - Identification of a senior-level manager or responsible person with the training, authority, and responsibility to have *complete control* as the coordinator during the incident or until public safety professionals assume control
  - Assignment of responsibility to the coordinator, who must have the ability to organize, deputize, and supervise
  - Assignment of trained response personnel or response teams, depending on the nature of the emergency or incident

  - Assignment of key management or other personnel to the command post
  - Assignment of key support personnel on and off premises
  - Assignment of public information officer or media spokesperson, who will also interact with victims and their families
  - Attention to employees affected by psychological or physical injuries, immediate and/or chronic, by trained human resources personnel
- Provide training for *all* employees on how to react to any serious incident that may occur on premises, including fire training and evacuation procedures.
- Maintain constant vigilance and routine inspection of fire equipment and hardware, first-aid equipment and supplies, and other safety equipment.
- Report all safety and security hazards or violations observed during routine inspections of the premises in a timely manner for correction.

Outside Emergency Support and Assistance

The Emergency Procedure Plan should include the following emergency services, organizations, groups, or trades along with the contact names, emergency telephone numbers, and the services they may provide:

- Local police/sheriff's department (county and state police, as may be requested by local departments)
- Local fire department (with mutual assistance agreements with other fire departments)
- Ambulance services/local hospitals
- Public utilities: electric, gas, telephone, water
- Emergency repair contractual services: alarm, glass, plumbing, electrical, heating and air conditioning, protection and preservation of property, clean-up, restoration services, etc.
- Suppliers of emergency equipment, particularly mechanical implements, fire hardware and first aid supplies
- Local Office of Emergency Management
- Local public officials
- Volunteer agencies: American Red Cross, Salvation Army, etc.
- Brokers and insurance carriers
- Local and national weather services

Protection of Vulnerable Areas

Identification of those areas or spaces that may be vulnerable to criminal trespass or a high probability of attack, particularly in areas open to the public:

- Secure all hardware, mechanical and maintenance rooms, and other important spaces.
- Secure drop ceilings and cabinets.
- Provide proper security for payroll and cash offices, accounts payable and receivable, and computer rooms and storage.
- Provide cessation procedures for cash registers and cash drawers in a retail environment.
- Provide proper trash receptacles.
- Provide proper furniture, cabinetry, and display cases.
- Remove any fixture that may easily conceal or contain a suspicious object or device.
- Maintain appropriate patrol and visual inspection by security personnel.

Security and Safety Measures

The level of security will depend on the business establishment and the strictness of the required control measures (e.g., limiting or denying access to the building or interior spaces).

- Strict lock and key control for perimeter and interior doors as may be necessary
- Access and presence on the premises only by identification
- Monitoring visitor movement via CCTV monitors, security officers, or the personal search of a visitor
- Storage of combustible materials according to fire codes, particularly near any heat source

Threat Analysis

Consider all bomb threats as serious and notify the police in every case. Response to the threat will depend on the analysis of the threat by the police and management.

Realizing that violence in the workplace usually cannot be foreseen, the following should be taken into consideration:

- Information received from employees concerning a disgruntled employee or former employee, or any indication that a present or former employee may be considered a threat at any time, must be considered and immediately acted upon.
- Any indication of an act of violence that may be or is in fact committed by an outsider must be acted upon immediately. This will include minor assaults or arguments upon employees by customers, clients, or visitors.

Incident Response

The foremost considerations for the initial responders to any disaster or emergency are the protection of life and attendance to the seriously injured. This may be considered the ***rescue*** or ***stabilization period*** of operation. During this procedure, the responder must try to ensure his or her own safety. Whatever the incident entails—fire, explosion, or a serious assault on the premises—immediate notification must be made to the proper public safety agencies for assistance.

During a fire or any type of explosion, evacuation of all inhabitants must be achieved as quickly and calmly as possible. Untrained personnel should not attempt to assist in the rescue of people trapped within a collapsed building. Leave such activity to trained emergency personnel.

In any biological or chemical exposure, the severity of injuries will depend on the nature of the agent and the extent of the exposure. If the spill or agent is contained, attempt to seal off and evacuate the area. Close off all ventilation and air-conditioning systems. Consider that the untrained responder to a biological or chemical exposure incident can offer no assistance that would be of value to the victims. Victims of these incidents will require immediate medical assistance. In the case of biological agents, assistance will also include monitoring and quarantine.

Other than biological or chemical incidents, if gas or vapors have entered the building, attempt to ventilate with caution by opening windows and doors, along with possible use of fans.

The command post coordinator must be able to respond to any requests for further assistance or equipment from the responders at the scene, but he or she may have to determine priorities, since limited resources and manpower will be available during the initial phases of the incident.

Timely response to requests for emergency apparatuses, trained specialists, and medical assistance from external agencies such as police, fire, and ambulance services must be handled by the coordinator. If possible, have someone at the command post compile and maintain a time record. A time record will be important when and if criminal charges arise from the incident. See the Glossary for the content required of a time record.

Upon the arrival of public safety officials, the coordinator will turn over command and cooperate fully in the protection of lives, meeting human needs, and achieving recovery. Professional rescue personnel and company supervisors should be aware of emergency responders and other employees, who may exhibit the following symptoms of stress: visual, auditory, and time distortions; tremors; tears; hysteria; nausea; hyperventilation; and weakness in the extremities.

- Attempt to establish a process for relief or breaks away from the scene of death, injury, or carnage so as to reduce the stress upon the responders. This area of relief should be away from the general confusion of the incident and as quiet as possible, with appropriate provision for rest and relaxation, along with an outside telephone for those employees who may wish to contact family.

- Following the reduction of the risk to human life and the protection of the injured, the protection or recovery of property may be considered the next priority.
- Subsequent investigation of the incident, whether criminal, accidental, or natural in origin, will be determined by the appropriate investigative agency and/or insurance carrier.

Determination of the Command Structure

Deployment of Arriving Resources

The following narrative applies to any incident that requires the establishment and manning of the command post because of a bombing, fire, or any catastrophe that has caused severe damage to the facility or death and injuries to the occupants.

The Command Post

Written policy as to the command structure that must occur in the event of a disaster or serious emergency must be made part of the Disaster Management Plan.

- The location of the command post or command central should be detailed.
- The manager, senior-level employee, or the person assigned as coordinator or director of the command post must be identified by position or name.
- If the initial person named as coordinator is not present when the incident occurs, subsequent authority to one or more persons to fill that position must be specified.
- The coordinator must have the authority and ability to completely control the incident—in effect, to ***organize***, ***deputize***, and ***supervise***. He or she must have the full support and confidence of the company administration to act.

The command post must have the necessary equipment to control the responders, to direct people and equipment to various areas as required, to request external assistance, etc. This equipment will include hand-held portable radios, public address systems, in-house firefighting equipment, first-aid equipment, and the necessary telephonic radio equipment for outside communication.

Remember that upon the arrival of police and fire units and depending on the circumstances and seriousness of the event, police and fire officials will take change of all emergency response responsibilities including the command post.

During the period when control is being transitioned to police and fire staffs, the director/coordinator will still be obligated to fulfill certain duties as detailed in the disaster management plan. However, cooperation with the public safety officials is essential.

The Responsibilities of the Command Post

In addition to those responsibilities of the command post detailed in the disaster plan, company management on or off premises must be informed.

- What management personnel are to be contacted in what type of emergency?
- When and how are they to be contacted?

If possible, maintain company functions in nonaffected areas. All information from the scene must be communicated to the command post upon request or in a timely manner.

Monitor Critical Areas for Safety

During any emergency, coordination and cooperation among company security and safety personnel and public safety authorities is imperative, if not by law then by common sense.

- Continue to respond to calls for service.
- Secure the perimeter.
- Coordinate traffic control around the facility.
- Assign security personnel, particularly to directly responding fire, police, and emergency services around and into the facility.
- Assign a security officer or a responsible person to direct public safety responders to a designated area and help maintain an open flow of traffic onto the facility grounds.
- Route the media and any responding family members to controlled designated areas outside of the perimeter.
- Depending on the nature and seriousness of the incident, be aware of when and who will announce the "all clear" signal for safe entry into the facility, when all personnel can return to their regular duties.
- Coordinate and cooperate with all public safety authorities

Immediate Post-Incident Activities

Once the command post is set up and manned, the appropriate deployment of outside forces can be achieved. Once the knowledge of the incident spreads and becomes known, the various public safety agencies will respond. The job of the command director is to advise and deploy those forces in an efficient and effective manner.

The proper notifications requesting assistance to those agencies not immediately aware of the incident—and the speed of deployment of those coming or already at the scene—is significant, since indecisive or poorly directed efforts can be the cause of confusion and uncertain assistance to those in need. As the incident unfolds, staging areas should be set up for arriving forces to await further instructions or to receive briefing for proper deployment to address new or continuing hazards. A staging area would most likely be set up by public safety authorities or in cooperation with loss prevention management and those authorities during prearrangement planning prior to an incident. With a staging area in place, newly arriving forces can also be better coordinated and matched with the proper equipment resources.

Access to the immediate area of the incident should be restricted, particularly if the area is to be considered a crime scene. This will include the media, which should be directed to a briefing area that is safe but in close proximity to the scene. A public information officer, appointed by the company, should be the only person that will make public comment. This spokesperson will defer to and assist public safety officials as they may require. He or she should also be privy to all facts and information as soon as possible so as to brief administrative company personnel and political officials.

Miscellaneous Actions of Concern Following the Incident

- Offer psychological, physiological, or social services as needed for anyone who was present at the time of the occurrence or who assisted in the emergency, as some of these people may have suffered some type of stressful reaction during or after the incident. The human resources department is the unit to coordinate this procedure.
- Once the precedence of saving lives and serving the injured is met, and the recovery of property is considered as complete as possible, the next priority is the restoration of the business operation.
- Facilitate the participation of assigned management and employees to the recovery team.
- Hire outside contractors for clean-up, recovery of property, and restoration.
- Assess and document the damage.
- Compile all incident and injury reports as well as any other pertinent forms, statements, or documentary evidence. These should be maintained as a complete packet for future reference or as may be required by law enforcement, insurance carriers, and company attorneys. (See incident report defined in the Glossary.)
- Notify all employees as to the status of the business enterprise.
- Return to normal business operations as soon as practical.

Procedures to Follow during a Civil Disorder

Civil Disorder Defined

Civil unrest can be defined as small gatherings of people secretly or openly discussing an incident or some type of grievance that becomes of great concern to the public at large. If the question continues to fester and grow among the populace, all that will be needed is a catalyst or agitator for the crowd to become a compelling force.

Civil disorder can be equated to a civil disruption or civil disturbance; a group or crowd milling about, drawn to a particular location because of some incident; or a form of protest against some sociopolitical issue. The trigger can be as simple as an impromptu sit-in, a parade, or some type of obstruction. The incident can be a cause as minor as an argument initiated by an auto accident or over a traffic citation, to an attack by a police officer on a young man involved in a minor incident. Subsequently, the general public becomes aware of the incident and recognizes or perceives the attack to be a case of police brutality. An act as simple as a dominant outspoken subject can easily cause the crowd's behavior to escalate into a civil disturbance, which can include property damage, assaults on people, and, ultimately, riot conditions.

Citizens not directly involved with the disturbance can be subject to inconveniences and disruptions in the social order, such as municipal services, transportation, access to various venues, and failure to obtain the necessities of life.

As a loss-prevention professional employed to protect an institution or business enterprise, the officer must respond to any civil unrest that could affect the business facility or its location:

- Upon determining that the disturbance will involve the facility, the local police should be asked to provide the necessary protection and preservation of the peace.
- Avoid provoking or in any way obstructing demonstrators.
- Avoid coming in contact with the demonstrators and, in particular, the agitators.
- If the situation escalates and it becomes impossible to continue normal routines, close the business. This will include the securing of the areas in and around the facility; lock all doors and safes and safeguard files, important records, and expensive equipment.
- Have all personnel and occupants remain inside and stay away from doors and windows.

If an intrusion is made by the demonstrators (we can assume at this point these demonstrators are now a mob or rioters), they have now committed a crime. If entrance is made for the purpose of causing sabotage, damage, larceny, or assault on the occupants, then the felony crime of burglary has also been perpetrated and a criminal trespass has been committed.

Security personnel may not encounter or attempt to arrest these perpetrators unless they have entered the premises and are causing damage to property or harm to the occupants. If damage is caused from outside the facility (e.g., rocks thrown against windows), security officers should not attempt to leave and chase the dissidents. The police should handle that type of situation.

> Check with your local and state criminal code to determine how much authority a security officer may possess off the property he or she is hired to protect.

State of Emergency

During any type of catastrophe or disaster, terrorist-related or not, the civil authorities may declare that a state of emergency exists, where certain functions of government may be suspended or ordered to implement emergency procedures. Certain civil liberties may be suspended, and martial law may be implemented, giving the military greater power to act. A state of war need not be a condition for this type of declaration to be made. Although the populace may accept this condition as a need for order and controlled activities for the good of all, there will be those who will criticize such methods as dictatorial and fascist, and not in keeping with a democratic society.

Workplace Violence: Mediation and Conflict Resolution

Recognizing the Potential for Workplace Violence

According to Larry Porte, a former Secret Service agent and presently a private-sector employee as manager and instructor of threat response and asset protection for the National Center for Manufacturing Sciences (www.ncms.org), "workplace violence is the product of an interaction among three factors":

1. The individual who takes violent action
2. The stimulus or triggering conditions that lead the person to see violence as a "way out"
3. A setting that facilitates or permits the violence—a setting in which there is a lack of intervention

Moreover, Porte states that perpetrators of violent acts usually have one of the following motives. He or she wants to:

1. Achieve notoriety or fame
2. Bring attention to a personal problem

3. Avenge a perceived wrong
4. End some personal pain and has a desire to be killed

In addition, NCMS is involved in safe security awareness fundamentals and violence educational programs.

Managing the Violent Antagonist

The following narrative is summarized from the About.com Web site edited by Susan M. Heathfield, concerning her guide to "Human Resources: Violence in the Workplace" (See http://humanresources.about.com/workplaceviolence). In order for the loss-prevention manager and the human resource manager (as the first line of defense) to begin to establish a policy and procedure regarding employee violence in the workplace, they should be aware of certain indicators and/or motives that can identify and hopefully resolve an act of violence.

The Art of Mediation and Conflict Resolution

Meet with the antagonists together. Let each briefly summarize his or her viewpoint, without interruption by the other party. This way it will be clear as to the disagreement or conflicting views. Intervene if either employee attacks the other employee physically or verbally.

Ask each participant to describe specific actions they would like to see the other party take that would, or could, resolve the differences.

You, as the supervisor, must own up to some of the responsibilities for helping the employees in the actions described in the second example noted above.

If the situation needs further exploration, ask each participant to additionally identify what the other employee can do to resolve the issue.

All participants are to discuss and commit to making the changes necessary to resolve the conflict. Commit the parties to treating each other with dignity and respect. It is okay to have reasonable disagreements over issues, but it is never okay to have personality conflicts that affect the workplace.

Let the antagonists know that you will not choose sides, that it is impossible for an outsider to the conflict to know the truth to the matter.

You expect the individuals to resolve the conflict proactively as adults. If they are unwilling to do so, advise them that you will be forced to take disciplinary action that can lead to dismissal for both parties.

Finally, assure both parties that you have faith in their ability to resolve their differences and to add to their successful contributions to your shared organization. Set a time to review the progress with both participants.

Actions to Avoid in Conflict Resolution

Do not avoid the conflict, hoping it will go away. It will always be there, cropping up with new disagreements or whenever stress increases in the work environment.

Do not meet separately with the participants. You will risk polarizing their positions; the employee will want to justify or convince you of the merits of his or her case, and cause suspicion of possible favoritism in the other employee.

Do not believe that the only people who are affected by the conflict are the participants. Every employee interacts with other employees. Friends will take sides, others will tread carefully in the presence of the antagonists, thereby creating a hostile work environment.

The Establishment of Policies and Procedures Concerning Violence in the Workplace

Policies

The company must establish a written policy under which all violence or potential violence by employees, and any indications or circumstances that may lead to on-site violence, will be taken seriously. In fact, the company should enforce a "zero tolerance" postion concerning any act of violence. All incidents of violence or attempted violence that may occur in or around the workplace shall be reported, as directed in the procedures dictated below. All employees must understand that any violent act against another person will be sufficient grounds for investigation, which may result in a police report, disciplinary action, and/or termination of employment. Note that sexual harassment may also be considered as an act of violence but should be dealt with under a separate policy and procedure.

Program Development

Once the company sets policy on workplace violence in writing, the initial step in developing a workplace violence prevention program is to design a *threat assessment team** (TAT) and define its operation and purpose. The TAT should be authorized and

* The organization that follows this policy and procedure example may wish to name the group in some other manner. For want of a better title, this chapter will identify this investigative group as the *threat assessment team*.

designated to assess the vulnerability to workplace violence and to reach agreement on the preventive actions that are to be taken. This team should be responsible for:*

- Recommending and implementing employee training programs on workplace violence
- The implementation of plans in responding to acts or threats of violence, and the responsibility of those employees required to react in those instances
- Communicating internally with employees

Procedures

Certain procedures must be formalized so as to establish how management shall react when advised of a violent incident or the possibility of such an incident.

1. Senior management or the administration of the company should establish a threat assessment team (TAT) whose duties shall include reviewing, investigating and recommending responses to all acts of violence. This team should consist of at least five managers and be separate from the safety committee, since the recommendations or sanctions offered may be immediate.
2. The responsibility for the authority, adjudication, and auditing of the workplace violence program should rest with the threat assessment team.
3. The threat assessment team should consist of the following on-site managers or their representatives:
   - The company/facility administrator or manager
   - The operations manager (as applicable)
   - The security or loss-prevention manager
   - The fire safety director (as applicable),
   - The human resources manager
   - The maintenance or engineering manager
   - The manager or supervisor of the person in question (if applicable)
   - The union representative of the person in question (if applicable)
   - A representative from legal and finance administration (as may be required for consultation)
4. The establishment of a written policy concerning workplace violence should indicate the following:
   - Zero tolerance for violence at work and implicit policy concerning the consequences, including any act that originates off premises and is carried into the workplace.
   - For any immediate and/or serious threat of violence in the workplace, the loss-prevention/security department should employ all legal and law

* Long Island Coalition for Workplace Violence Awareness and Prevention, *Workplace Violence Awareness and Prevention,* February 1996.

enforcement sources for assistance. If the threat is a danger or an act of violence by and against an employee, appropriate authorities should be advised and necessary action taken in an attempt to thwart the act.

- Development of various mechanisms for the reporting of violence, the threat of violence, or the possibility of violence.
- The promotion of open communication and input from employees and supervisors regarding stress in the workplace that may affect absences, poor work habits or production, or as an indication of possible violence.
- The written documentation of poor performance, erratic and abnormal behavior.
- Statistical information about past violent behavior occurring in the various groups: employees, customers, clients, and visitors.
- Periodic reviews of such statistical information regarding violent incidents so as to identify the hazardous areas of employment, possible reasons for violence that may occur, and preventive measures to reduce or eliminate such incidents.
- Investigation the occurrence.
- Investigation the seriousness of the threat, ability to carry out the threat, and the action necessary to prevent the threat from occurring.
- The determination as to who on the team will be responsible for:
  - Immediate care for the employee/victim
  - Follow-up care such as stress debriefing sessions with the victim, coworkers, and possibly the families
  - If the incident involves employees only, the follow-up and enforcement of the sanctions and compliance as noted below

5. Human resources personnel and all members of the loss-prevention/security department must receive adequate professional training in all aspects of violence in the workplace.
6. All managers and supervisors should also have access to this type of training in evaluation, reporting, and interaction concerning acts of violence, threats of violence, and violent employees.
7. A manager or supervisor who becomes aware of any act of violence, threat of violence, or potential violence shall immediately report same to the loss-prevention manager, human resources manager, or the facility manager. The loss-prevention manager shall conduct an immediate inquiry to determine whether the incident requires immediate intervention or a report to the threat assessment team.
8. An employee who becomes aware of an act of violence or potential violence shall immediately report it to his/her supervisor or the manager of human resources.
9. If the act appears to be an immediate threat, the employee shall report the threat to the loss-prevention manager or security personnel immediately.

10. The loss-prevention manager will be notified of any act of violence and, depending on the circumstances and after a thorough investigation, inform the local law enforcement authorities of any act that constitutes a violation of public law. If the loss-prevention manager is unable or unauthorized to act as the complainant, the victim may be assisted in originating a police report on the incident.
11. All reports by an employee of violence, threats of violence, or potential violence shall be kept confidential, with only the members of the threat assessment team and security personnel *involved* aware of an informant's identity. If the identity of the informant is required for any police or criminal justice action, such identification shall be made by the police as required by law.
12. All oral discussions, written reports, or records identifying the informant, the offender or apparent offender, or the incident itself shall be strictly controlled for confidentiality, unless exposure is required by law.
13. The threat assessment team shall convene as soon as practical after any member of the team receives a report of violence or suspected violence.
14. When the threat assessment team has been convened, it shall review and evaluate the act or threat of violence. If required and appropriate, the team may wish to seek the advice of the police, a mental health professional, or a behavioral specialist who has a background in workplace violence.
15. The threat assessment team shall establish certain procedures to be followed for each incident. This may include the development of a strategy for dealing with particular acts of violence:
    - The assignment of appropriate personnel to implement such procedures
    - The reporting to the local law enforcement authorities as may be required by law or company policy, depending on the type of incident or occurrence
    - The assignment of certain personnel to maintain statistical data on all violent situations
    - Delineation of policy regarding minor repetitive acts by an offender and the consequent sanctions
    - The assignment of certain personnel to control and determine that sanctions or compliance concerning serious violent behavior are being complied with
    - The confidential maintenance of all records pertaining to the incident
    - Informing the victim or possible victim that such act or threat has been reported to the threat assessment team and an investigation is in progress; the victim shall also be kept informed of all results
    - Consideration of contractual agreements with union personnel regarding suspensions without pay, representation, or other sanctions

Sanctions and Compliance

There must be a strict policy of zero tolerance when an act or threat is of such gravity that termination of employment must be an immediate reaction by management. This would include any act concerning a threat or an attempt of suicide, a physical altercation where an injury is suffered and the intentional destruction of company or personal property.

Of course, proper investigative procedures must take place prior to any termination process. The possibility of police involvement may also be a consideration. Necessary reports should be submitted to the threat assessment team describing the incident and the reasons for the termination of the subject or subjects involved. If the threat assessment team finds after deliberation that an incident under investigation warrants disciplinary action against the offender, it may pursue the following course of action:

- All acts of violence may be considered grounds for suspension without pay and/or termination of employment.
- A psychological assessment, a fitness evaluation, or counseling can be offered to the subject on a voluntary basis or as a condition for continuation of employment.

Guidelines for Evaluating Bomb Threat Credibility

Threat Analysis

On many occasions, a terrorist will accompany an act by a threat or a claim of credit. There can be logical explanations for this type of notification, and one version is that the caller is not in fact a terrorist. The subject of the threat, whether it is a person, company, or institution, must determine the genuineness of such a claim and consider the following:

1. The caller is in fact a terrorist or is affiliated with the terrorist group taking credit for the act, which is created to inflict death, injury, and damage.
2. The caller may or may not be a terrorist, but has definite knowledge of the event. The caller's action may be made to eliminate or lessen injuries and damage, or it may be motivated by conscience.
3. The call is bogus and the caller wishes to create an atmosphere of anxiety at the targeted facility and its populace as well.
4. The caller is not a terrorist, but in fact is deranged in some manner, or has a grudge or hostility against the company targeted.

The most frequently used method of notification is by telephone, before or after the act. However, recently many terrorist organizations have made claims after incidents by using the media to publicize their particular viewpoint or propaganda to as large an audience as possible.

Threat Credibility

The determination of the credibility of a threat frequently involves a decision made by a number of persons such as security, building ownership or management, and law enforcement personnel who will base that decision upon the consideration of many factors.

The following guidelines cover common factors that should be considered when evaluating the believability of a bomb threat. There may be others directly related to a particular situation. However, each is worthy of consideration, and as a general rule, the more indicators present, the more credible is the threat.

Bomb Threat Evaluation Guidelines

The following *indicators* and *considerations* for the evaluation of a bomb threat are described in some detail:

Indicator 1: The quality and effectiveness of the facility's access control systems.

Consideration: If a bomb threat is received at a facility that has good perimeter protection, the chances that an outside group or individual placed a bomb in a work area are limited. Conversely, if there is no or only a limited access-control system in place, or if the building is open to the public, the placing of a bomb could occur more easily.

Indicator 2: Did the caller of the bomb threat display specific and definite knowledge of the facility? Did the caller provide details as to the time of the explosion, location, and reason the bomb was placed? Did the caller sound intoxicated; was there laughter in the background, or other signs that the call is a prank or hoax?

Consideration: The more specific the threat, the higher is the level of credibility. More credence should be given to the caller who displays first-hand knowledge of the physical layout of the building and the operating function of the facility, and who is willing to provide greater details as to when the bomb will go off.

Indicator 3: Has there been unfavorable publicity against the company or its industry? Is the company involved in an activity that is highly controversial?

Consideration: Unfavorable publicity to a company may focus unwarranted attention on its operations and may make the facility vulnerable to bomb threats that prove to be false. On the other hand, involvement in highly controversial issues, or emotional and volatile situations, may require the threat to be taken more seriously.

Indicator 4: Was the time limit given realistic?

Consideration: Since the intent of most serious bombers who call in a threat is to avoid hurting people who may be sympathetic to their cause, a serious threat is usually timed to enable innocent people to leave the area without injury. However, if the threat has political, cultural, or religious overtones, the probability of a bomb causing catastrophic destruction, serious injury, and death will usually take place immediately or very soon after the call, if a call is made.

Indicator 5: Was there a threat of multiple bombings?

Consideration: Experience has shown that threats of multiple bombings within a single facility will more likely be false because of the difficulty in coordinating such an undertaking, although multiple bombings are possible, and must be considered.

Indicator 6: Have there been previous terrorist threats?

Consideration: If a terrorist group has recently made declarations or threats against an institution or a company, particularly if that company has international holdings, any subsequent bomb threat should be seriously considered in conjunction with other indicators. Similarly, if the details of the threat fit other actual incidents, the threat has more credibility.

Indicator 7: Have previous bomb threats resulted in evacuation?

Consideration: Media publicity of a bomb threat incident often fosters more bomb threats. Threats may be made to try to halt company operations again and again. If there is no history of an evacuation and if the bomb threats are isolated incidents, the credibility of the threats increases.

Bomb Threats

Bomb threat procedures and subsequent actions may require different approaches to the danger. Because this area of concern is so important, more than one threat procedure is detailed and offered to the reader. The loss-prevention manager or security officer may make use of each procedure alone or incorporate both into one plan. The two plans are:

Bomb Threat Procedure
Bomb Threats and Search Techniques

Both procedures are covered in the following sections, along with a Bomb Report Checklist.

Bomb Threat Procedure

Very few bomb threats ever materialize, but proper preparation and planning for threats can provide protection of life and property if the situation is in fact real. We cannot overlook the fact that the threat of terrorism is becoming more likely each day. This is particularly true for aggression against large national or international corporations. Therefore, we must:

Take all bomb threats seriously—never disregard a threat as a hoax.

The information contained herein does not cover every bomb threat incident, but will help to direct the efforts of loss-prevention and building personnel in the response to such a situation.

The protection of life and limb is the primary consideration for security personnel. Following that, the protection of property is the next concern. In addition, there is the importance of immediately relaying information to the proper authorities, and the initiation of accepted procedures in a manner that does not cause panic, concern, or harm among employees and visitors.

When a bomb threat is received, whether by telephone or by written communication, all facts regarding the threat and incident should be recorded on a proper form (see Bomb Report Checklist in the following section). During the telephone conversation and immediately following its conclusion, the checklist should be completed with as much information as possible. Immediately thereafter, the site security supervisor should be notified. The police or local law enforcement authorities must be notified as soon as practical.

If an explosion occurs, immediately notify the fire department and the police department.

The Fire or Emergency Procedure Plan for evacuation of the facility should be activated.

If the facility has a written Emergency Procedure Plan, the plan should be placed into effect. Certain notifications must be made, key company and building personnel must be immediately advised, and certain actions must be followed. Public safety personnel will take charge of the incident upon arrival at the scene.

An incident report should be compiled after the incident is concluded. A time log of all occurrences should also be maintained. This time log will include the time of the report, times of all notifications made, time of arrival of public safety personnel, and when all managers, supervisors, and persons of concern arrive at the scene. The Bomb Report Checklist form, along with the time log, should be attached to and made part of the incident report.

Upon the receipt of a written communication advising of a bomb threat or a suspicious letter or package that may contain an explosive device, do not attempt to handle the item more than necessary so as not to obliterate any forensic evidence or cause an activation of the device. Again, advise security management and the police as soon as practical.

Whether or not the time of detonation for the device is known based on the information received, and whether or not the threat is believed to be bogus, *the police must be notified*. Depending on the facility's emergency plan based upon company policy and procedure, management may require an immediate evacuation of all personnel and visitors, or may elect to wait for the arrival of the police before any action is taken.

When a caller makes a bomb threat, as much information as possible should be gathered from the caller. Study the company's Bomb Report Checklist; memorize key points. Keep these forms near telephones that take incoming calls. The information contained on this form will be the basis for how the police will conduct their actions.

Regarding buildings containing children and students, the general procedure is that, when a bomb threat is received, the police will evacuate any school, playground, or public area where children might be present.

Depending on a particular police department's policy, one of three actions will be taken. The police will determine how serious the threat is based on the information noted on the Bomb Report Checklist form, the interview with the receiver of the call, other incidents that may have occurred in the area, and police intelligence information. Based upon their experience and on the facts in this case, the police will advise company management to:

1. Immediately evacuate the premises
2. Recommend that the premises be evacuated, but leave the decision to the company
3. Leave the evacuation decision to management without any recommendations

In any event, a search for the device will be conducted whether the facility is occupied or not. Although the sight of police and other emergency personnel within the facility will give rise to the concerns of those within, attempt to keep the incident as low key as possible.

Prevent the disruption of normal business operations if possible, but do not avoid answering questions asked by occupants or visitors, so that individuals can decide whether to leave the premises until the threat has passed. When there is a possibility of danger, do not deny that an emergency exists in an effort to abate fear and place people at ease. Doing so exposes the employee and the company to possible liability. Be truthful, but if top management requests that you not explain what is happening if asked, conduct yourself accordingly, but make appropriate notations in the incident report.

Avoid undue panic or hysteria.

Remember, the individual facility manager or the loss-prevention manager (if given that authority) will decide whether to evacuate (unless directed by the police) or to notify employees of the bomb threat. If company management decides to inform employees of the threat or wishes to evacuate, security and building personnel will cooperate with management's wishes for an orderly evacuation of the facility. Other than to advise or recommend movement from the building, security officers will not demand or in any way conduct an evacuation unless they believe that the threat is so immediate that there is no time to make the necessary notifications and consultations.

Once the police have responded to the notification of the bomb threat, they will have complete control of the scene. Let the professionals in public safety handle the situation. An emergency plan should note by whom and where interior and exterior searches will be made. Generally, a police officer will accompany a security officer, maintenance employee, or other employee who has keys and access to all parts of the facility. A police supervisor or a site security manager should set up and direct the search teams so that the search will be completed in a timely manner with no duplication. All public, employee, and fire stairways and stairwells will be thoroughly searched. Search also all restrooms, stalls, unlocked and locked closets, mechanical rooms, public and private lockers, locker rooms, trash and garbage receptacles (interior and exterior), under displays and desks, and anywhere a suspicious package can be left. Consider any paper bag, briefcase, attaché case, knapsack, backpack, handbag, or any other item that appears out of place as a possible explosive device. With present plastic explosives and technology, a bomb or incendiary device could be as small as a package of cigarettes. If you have doubts, report an item as suspicious.

When a suspicious package is found, do not touch it. Call for a police officer to respond if not already present. Evacuate the area and safeguard the scene.

Do not use a portable radio to transmit a message. All pagers and radios should be turned off and not used after a bomb threat is received. Any transmission of this equipment could possibly activate a device by electronic signal.

Keep calm in all emergency situations. Think before you act. Your actions and demeanor will affect others.

Be guided by the police or other public safety personnel. Remember that the safety and protection of life and limb are the first considerations of all security personnel.

After a search has been completed and the facility is considered safe by the police, and if in fact an evacuation had taken place, reentry by employees and visitors will be authorized only by the police or other public safety officials.

The information contained herein covers basic procedure. Let the professional public safety personnel handle and direct any incident or serious threat where facts and circumstances may alter the emergency situation, and where *only they* have the legal authority to act under emergency public safety conditions.

Clarification

Bomb Search Participants

During a bomb search, the question will arise as to who will participate. Police officers and loss prevention/security and safety officers are emergency response personnel and are obligated by their job descriptions to assist. However, other employees of a facility who possess keys and have access to all spaces within the building may be requested to assist. These employees may include managers, maintenance, and/or engineering personnel.

It has been recognized that some of these employees may feel that they are not emergency providers, particularly not in the case of bomb searches. In addition, they may indicate that they are not police offices or security and safety personnel and that their job description does not entail functions or responsibilities of this type. You should be aware that it has been held that anyone has the right to refuse to put himself or herself in possible danger. If a police, private security, or safety officer refuses to obey a direct order from a superior, summary action may be taken against that officer. Other employees who want to withdraw will not be held liable for their refusal. No action, dismissal or otherwise can be taken against them. Their actions must be completely voluntary, and for accountability and liability, all participating employees must be listed in the incident report.

BOMB REPORT CHECKLIST

GENERAL TELEPHONE INSTRUCTIONS UPON RECEIVING A BOMB THREAT

LISTEN – DO NOT INTERRUPT THE CALLER. BE CALM. BE COURTEOUS.
TAPE RECORD THE CONVERSATION IF POSSIBLE.
ATTEMPT TO WRITE OUT THE COMPLETE MESSAGE GIVEN BY THE CALLER.
PROLONG THE CONVERSATION.

DETERMINE AND NOTE AS MUCH OF THE FOLLOWING INFORMATION AS POSSIBLE

EXACT WORDS OF CALLER: ______________________________

____

____

____

TODAY'S
DATE: __________ ORIGIN OF CALL: Local ___ Long Distance ___ Booth ___ Within Bldg.___ Unknown ___

TIME OF CALL: __________ CALLERS IDENTITY: Male ___ Female ___ Juvenile ___ Approx. age: ___

BOMB FACTS

Pretend difficulty hearing the conversation. Keep caller talking. If caller seems agreeable to further conversation, ask questions such as the following—but do not anger the caller.

When will the bomb go off? __________ Hour? __________ How much time remaining?__________

Where is it located? ____________________ What area of the bldg.? ____________________

How do you know so much about the bomb? ______________________________

Why are you doing this? ______________________________

Where are you now? ______________________________

Who are you? ______________________________

What is your name/address? ______________________________

IF THE BUILDING IS OCCUPIED, INFORM THE CALLER THAT A DETONATION COULD CAUSE INJURY OR DEATH

CHARACTERISTICS OF CALLER Check off all that apply

VOICE	SPEECH	ACCENT	LANGUAGE	MANNER
___ Familiar	___ Fast	___ Local	___ Excellent	___ Calm
___ High pitched	___ Slow	___ Not local	___ Good	___ Rational
___ Deep	___ Distinct	___ Foreign	___ Fair	___ Laughing
___ Incoherent	___ Coherent	Type?______	___ Poor	___ Intoxicated?
___ Raspy	___ Stutter	___ Race?	___ Deliberate	
___ Loud	___ Slurred	___ Region?	___ Foul, obscene	
___ Soft	___ Righteous	______________	___ Other	
___ Pleasant	___ Nasal		___ Angry	
	___ Distorted			
	___ Irrational			
	___ Lisp			
	___ Emotional			
	___ Other?			

FAMILIARITY WITH THIS FACILITY?

___ Much: Explain ____________________

___ Some: ____________________

___ None apparent

BACKGROUND NOISE

__ Factory machines	__ Office machines	__ Animals	POLICE NOTIFIED? Yes ____ No ____
__ Bedlam	__ Music	__ Mixed/noisy	Time Ntfd. ________ Time of Arrival ____________
__ Street traffic	__ Trains	__ Airplanes	
__ Voices	__ Party atmosphere	__ Quiet	**POLICE SUPERVISOR @ SCENE** ____________

PERSON REPORTING ______________________________ Signature ______________________________

If needed, use other side of this form for further information.
NOTIFY YOUR SUPERVISOR IMMEDIATELY AND FOLLOW THE EMERGENCY PROCEDURE PLAN.

Give original copy to police if requested.
Attach this form or a copy to the Incident Report, where further information on the incident will be recorded.

Bomb Threats and Search Techniques

If a telephone call regarding a bomb threat is made to an institution or business, there only two reasonable explanations.

1. The caller has definite knowledge as to when, where, and how the explosion will occur. The caller, who may or may not be the person who placed the bomb, wishes to minimize death, injury, and destruction as much as possible.
2. The caller wants to create an atmosphere of panic, fear, and anxiety, which will result in a disruption of all activities at the place where the bomb is deployed.

In either case, the caller may be a deranged person or a disgruntled individual who has a grievance against the occupants or the building targeted.

However, bear in mind that, if there is no warning of an explosive device, we can assume that the act is of a terrorist origin. In this case, after the activation of the bomb, the perpetrator may advise the target or the public, through the media, who did the act and why it was done.

Preparation

1. Establish a list of those personnel who will conduct a search for a bomb or suspicious object. Depending on the facility, this may include all security officers, safety officers, fireguards, members of the fire brigade, and all maintenance personnel. Proper training in search procedures will be conducted routinely for those personnel.

2. Train all security and maintenance personnel to be alert to suspicious-looking or unfamiliar persons or objects that appear to be out of place.
3. Require security and maintenance personnel to make periodic checks of all restrooms, locker rooms, stairwells, and any other area where a person or object may be unusual.
4. Lock all electrical rooms, generator rooms, mechanical rooms, elevator mechanical rooms, closet and storage rooms at all times, with access only by authorized personnel.
5. Establish strict procedures and controls for the inspection of packages and materials entering critical areas.
6. Develop a strict policy for identifying personnel who have authorized access to critical areas.
7. Determine from the local law enforcement authorities whether a bomb disposal unit is available, and under what conditions it may be deployed. Will the unit be involved in the search for the bomb, or respond only when a suspicious device is found and located?
8. Arrange, if possible, for the local police precinct and fire companies to visit routinely, with your staff, to familiarize themselves with floor plans, building layout, and areas of concern.
9. Determine if there is adequate protection for classified documents, proprietary information, and other records essential to the operation of the institution or business.
10. Determine if there is adequate protection for computers and peripheral equipment. Trained teams of personnel must be available on site for immediate response to any computer problem or attack on the system.
11. Train all telephone operators and receptionists in how to handle an incoming bomb threat, and the notifications to be made thereafter.

Evacuation procedures for various incidents should include routine training of all company personnel.

Evacuation

Regard evacuation as the most important action to be taken immediately after a bomb threat notification. Management must protect all occupants within the building and make a decision promptly. If an evacuation is conducted and the threat turns out to be a hoax, the result can be a disruption of the normal activity and a reduction of production at least for a short period. Because it is almost impossible to determine immediately whether a threat is bogus or not, and because today there are more threats that materialize, the decision for evacuation may be indisputable.

Consider that the method of the evacuation notification should be the same as the business would conduct for a fire emergency. To use a different signal, or to advise otherwise of a bomb threat, would increase the confusion, panic, and anxiety of the occupants, causing a rush to exit the building, which could cause serious injuries.

If the business is located in an office building with more than one tenant, emergency and evacuation procedures must be set up and controlled by the building manager. Cooperation of all tenants in these matters is most important.

Realize that any evacuation will cause an inconvenience to the public, particularly in public buildings, and company personnel can expect negative comments, no matter how serious an event might be.

*Bomb Search Techniques**

1. A plan for a bomb threat procedure must be written and made part of the Emergency Procedure Plan. It should detail when a command center is to be formed; where it should be located; who should staff the center, including the supervisor or manager in charge; and the requisite duties, responsibilities, and controls. This command center will be the central locus of operations during the emergency.
2. If a suspicious package or object is discovered, do not touch or handle. The location and description should be reported to the person or police officer designated to receive this information.
3. The removal and/or disarming of the bomb device must be left to the professionals in explosive ordinance disposal (the police or other civil/military authorities).
4. If the danger zone of the device is located, the area should be blocked off or barricaded with a clear zone of 300 feet until the area is deemed safe.
5. All gas and fuel lines must be turned off, along with all electrical power at the main switch, with consideration of enough light to conduct a decent search.
6. During a search of the building, rapid communication is of utmost importance. Attempt to use existing building telephones and public address systems. All handheld radios or cell phones must be turned off. Their use can be dangerous in that any radio transmission energy can cause a premature detonation of an electric or electronic initiator (e.g., a blasting cap).
7. With the assistance of the police, controls must be established to prevent unauthorized access or reentry to the building until the danger has passed.

---

* All search techniques, room searches, and outside procedures have been summarized from "Bomb Threats and Search Techniques," Publication ATF P 7550.2, Bureau of Alcohol, Tobacco and Firearms, Department of the Treasury, Washington, DC.

8. All evacuated persons must be placed at a safe distance from the building in order to protect them from any debris or flying objects if an explosion does in fact occur.
9. During the emergency, medical teams should be on standby within a safe distance from the building in case of an explosion or other circumstances.
10. Fire brigade personnel should also be on standby, ready to operate fire equipment if necessary. The municipal fire department may already be advised and be on site as a precaution.
11. Finally, pre-emergency plans should include temporary locations for the business operation in the event that a bomb threat does materialize and the building is determined to be unsafe.

Room Searches

The following technique is based on the use of a two-person search team. There are many possible variations in searching a room. The following contains only the basic techniques. The loss-prevention manager may wish to contact local civil authorities for other search procedures that may be available.

1. ***First team action—listening***: Upon first entering a room to be searched, the two-person team should move to various parts of the room and stand quietly, with their eyes shut, and listen for a clockwork device. Frequently, a clockwork mechanism can be quickly detected without the use of special equipment. If no clockwork mechanism is detected, the team is now aware of the background noise level within the room itself. Background noise or sound that is transferred may be disturbing during any search. Any ticking sound or noise that cannot be located can cause one to be unnerved. Sounds may come from an unbalanced air conditioner or ceiling fan or a dripping sink down the hall. Sounds may travel through air ducts, along water pipes, particularly hot water pipes, and noise from the street outside of the building.
2. ***Second team action—division of the room and selection of search height***: The person in charge of this search team should look around the room and determine how the room is to be divided for the search and to what height the first search sweep will take.

   Dividing the room: The room should be divided into two equal parts or as nearly as equal as possible. This division should be based on the number and type of objects in the room, and not the size of the room. Draw an imaginary line along the wall, e.g., the bottom edge of a window on one side of the room to the bottom edge of a window on the other side of the room. It is suggested that an electronic or medical stethoscope be used during any room search.

   The first searching height: This first searching height covers all items up to hip height. Look at the furniture or other objects in the room and determine the average height of the majority of the items resting on the floor. In an average room, this will include tables, desks, chairs, etc., and may include cabinets and air conditioners along the wall up to hip height.

3. ***First room-searching sweep***: Once the searching height has been selected, both searchers will go to a particular point in the room and, back to back, start to search across the room, checking all items resting on the floor, around the wall area of the room, and attached to the wall, until they meet again at the other side. At this point, a wall sweep will have been completed. They should then make a search of all items within the room up to hip height. This first searching sweep will consume most of the time and effort.
4. ***Second room-searching sweep***: The person in charge of the sweep will again view the objects in the room and determine the height of the second searching sweep. This will usually cover the height between the hip and the chin or the top of the head. The two searchers will return to their starting point and repeat the searching techniques at the second search sweep height, covering pictures on the walls, cabinets, bookcases attached to the wall, tall lamps, etc.
5. ***Third room-searching sweep***: Upon the completion of the second sweep, the person in charge will determine the next searching height. This is usually from the chin or top of the head up to the ceiling. This sweep will include high-mounted air conditioning or heating ducts, hanging light fixtures, etc.
6. ***Fourth room-searching sweep***: If the room has a false or suspended ceiling, then the fourth sweep will investigate this area. This will include flush- or ceiling-mounted fixtures, ventilation ducts, sound or speaker systems, electrical wiring, and fire sprinkler heads.

Upon completion of the room search, a sign or marker should be posted noting "Search Completed." If a sign is not practical, police tape should be placed across the door and doorjamb of the room to advise other searchers that this room has been searched and cleared.

This basic search technique can be expanded and used in other areas such as a convention hall or an airport terminal.

Searching Outside Building Areas

When searching outside areas and the building perimeter, pay particular attention to trash receptacles, garbage cans, dumpsters, incinerators, manholes, and street drainage systems. Include a check of parked automobiles and trucks, particularly if the parking areas are close to or part of the business complex.

Suspicious Object Located

If an object is found and determined to be suspicious, it is not to be moved, touched, or in any way disturbed. Bombs are made to explode, and there are no absolutely safe methods in handling them.

Search teams are only to be used for searches only, and are to report any suspicious object to the professionals, who will determine if the object is in fact an explosive device and then attempt to properly and safely disarm or dispose of it. If not already in command of the incident, the professional bomb disposal unit will control the emergency completely at this point and will determine how to safeguard the scene and conduct the disposal.

In conclusion, to search an area, a search team should:

1. Divide the area and select a search height.
2. Start from the bottom and work up.
3. Start back-to-back and work toward each other.
4. Go around the walls then to the center of the room.

Encourage the use of common sense and logic in any search.

The Fire Safety Plan Summarized

Purpose

The Fire Safety Plan shall be written and disseminated by management to establish a method of systematic, safe, swift, and orderly evacuation from a building by its occupants in the event of a fire or other emergency. In reality, the purpose of this guide is to outline the duties and responsibilities of the building staff and the tenants in a fire emergency. Further, it details the assignment of fire wardens, members of the fire brigade and other essential tasks. Moreover, if the facility in question is a high-rise building and consists of more than one tenant on one or more floors, the management of each tenant will be guided as noted in the following narrative. All tenants must comply with this directive. Although the plan described here relates to a vacant building with multiple floors, it can easily be adapted to a single tenancy on one or more floors.

These procedures are to be attached to and made part of the building's Emergency Preparedness Plan.

The following table of bomb search systems, provided by the ATF, is another useful resource.

BOMB SEARCH SYSTEMS

	Search by Supervisors	*Advantages*	*Disadvantages*	*Thoroughness*
Supervisory	Best for covert search. Poor for thoroughness. Poor for morale detected.	1. Covert. 2. Fairly rapid. 3. Loss of working time of Supervisor only.	1. Unfamiliarity with many areas 2. Will not look in dirty places. 3. Covert search is difficult to maintain. 4. Generally results in search of obvious areas, not hard to reach ones. 5. Violation of privacy problems. 6. Danger to unevacuated workers.	50-65%
	Search by Occupants	Advantages	Disadvantages	Thoroughness
Occupant	Best for speed of search. Good for thoroughness. Good for morale (with confidence in training)	1. Rapid. 2. No privacy violation problem. 3. Loss of work time for shorter period of time than for evacuation. 4. Personal concern for own safety leads to good search. 5. Personal conducting search are familiar with area.	1. Requires training of entire work force. 2. Requires several practical training exercises. 3. Danger to unevacuated workers.	80-90%
	Search by Trained Team	Advantages	Disadvantages	Thoroughness
TEAM	Best for safety Best for thoroughness Best for morale. Poor for lost work time.	1. Thorough. 2. No danger to evacuated workers	1. Loss of production time. 2. Very slow operation. 3. Requires comprehensive training and practice. 3. Workers feel that company cares for their safety. 4. Privacy violation problems.	90-100%

Note: Since the institution of Homeland Security, the Bureau of Alcohol, Tobacco and Firearms (ATF) advises that explosives, bomb threat and detection resources are no longer available online. Publication numbers and titles are offered, but one must send a written request on one's official letterhead to obtain copies of any publications produced by ATF.

The Fire Safety Director

The fire safety director (FSD) shall have complete authority to organize and assign building service personnel to the fire brigade. This person is also responsible for the state of firefighting readiness, fire and evacuation drills, and the designation and training of a fire warden and deputy fire wardens for each floor of the building in accordance with local fire department rules. Further, the FSD is responsible to see that periodic inspections and examinations of all firefighting duties and that assignments for building service and tenant personnel are carried out. During a fire emergency, the FSD shall supervise the fire command station and coordinate firefighting strategies and evacuation.

Generally, in a high-rise facility that contains various tenants on each floor or within each floor, a fire safety director is employed for the entire building. If there is no designation for a position of a fire safety director, the loss-prevention manager must assume these duties.

Assignment of a Fire Warden and Deputy Fire Warden

1. A responsible and dependable tenant employee shall be assigned as a fire warden or a deputy fire warden.
2. Such fire warden shall be provided for each tenancy. If a tenant exceeds 7,500 square feet of occupied space, a deputy fire warden shall be assigned to each 7,000 square feet or part thereof.
3. Six deputy fire wardens, at a minimum, are required to be assigned to each full floor.
4. A sufficient number of tenant employees shall be assigned as searchers. These employees shall determine that all offices, restrooms, lavatories, break rooms, closets and storerooms, and other spaces are devoid of personnel during an evacuation.
5. The names of all employees of the tenant so assigned as fire warden, deputy fire warden(s), and searchers shall be submitted to the building fire safety director on a fire warden list provided by the building owner or manager. This list shall be updated periodically.
6. A chart designating the floor fire warden, deputy fire warden(s), and searchers shall be prepared, posted in a conspicuous location for each tenancy, and updated periodically.

Duties of the Fire Warden

1. Be thoroughly familiar with the Fire Safety Plan and the operation of any fire alarm system.
2. Be thoroughly familiar with all fire exits on the assigned floor, the number of occupants, and the formulated traffic pattern for evacuation through emergency exits.

3. Conduct a daily inspection and examination of all fire doors and fire stairways to verify that fire doors are kept closed and that both fire doors and stairways are free of obstructions and maintained as required. Passageways, corridors, and aisles shall also be kept free of obstructions.
4. Determine that the required fire exit signs and lighting are in good operating condition.
5. Enforce fire drills and the testing of communication and alarm systems; all tenants must comply with the tests when conducted by the building fire safety director.
6. Compile a list of all personnel with physical disabilities who cannot make use of the stairways unaided; submit this list on a form provided by the building fire safety director.

Responsibilities of the Building Staff or Company Management

Fire safety director or a deputy: In a fire emergency, shall report to the security desk or fire command station to supervise, coordinate, and direct that:
- The fire department is notified and, upon arrival, that the fire officers are advised of the fire conditions and given all of the emergency keys
- All elevators are directed to the ground floor and that proper evacuation procedures are followed

General or building manager: In a fire emergency, shall report to the security desk or fire command location to assist the fire safety director as may be needed.

Operations manager or deputy to the building manager: In a fire emergency, shall:
- Assist the general/building manager in carrying out his or her duties
- Assume the full duties and responsibilities in the absence of the general/building manager

Fire brigade: In a fire emergency, members of the fire brigade shall respond to the fire location in an attempt to fight and control the fire, if possible, and to assist the wardens in evacuating personnel from the fire area.

Evacuation brigade: In a fire emergency:
- Designated members of the building staff shall be appointed as members of the evacuation brigade
- Upon notification of a fire alarm, various designated members shall report to
  - The fire command station
  - Main pump room or sprinkler room
  - The emergency generator room
  - The security desk, so as to communicate with floor wardens, tenant wardens, and tenants
  - Assist the fire department as to location and conditions of the fire

Responsibilities of the Tenant

Unless authorized and directed by building management, a company that occupies one or more floors in a high-rise building as a tenant should adhere to the following provisions:

Floor Wardens

If there is more than one company tenant located on one floor of a high-rise building, one floor warden should be assigned and agreed upon to act as the warden for that floor, and all tenants must comply with his or her directions during a fire emergency. In addition, the floor warden shall direct the evacuation of the floor in accordance with the directions received from the fire command station and the following guidelines:

- Utilize tenant wardens to ensure that all occupants of the floor are notified of the fire and to immediately initiate the evacuation plan.
- Know all personnel who are disabled and will be in need of assistance during an evacuation and maintain an up-to-date list of all personnel who fall into this category.
- Direct all personnel to the safest stairway based on information received from fire officers or the security desk and personal observation of the environment in the stairwell.

Tenant Wardens

The tenant warden shall assist the floor warden in the implementation of his or her duties and responsibilities, and ensure that all occupants are evacuated within his or her tenant spaces. In the absence of the floor warden, the tenant warden shall assume the duties and responsibilities of that position.

Assistant Tenant Wardens

The assistant tenant warden shall assist the tenant warden in the implementation of his or her duties and responsibilities, and assume the duties and responsibilities of a tenant warden in his or her absence.

Searchers

Searchers shall be assigned to each floor, and their duties and responsibilities are to check all offices, conference rooms, restrooms, break rooms, storage rooms, reception and lobby areas, and any other spaces or remote areas so as to notify, advise, and ensure that all personnel and visitors are evacuated.

Monitors

Monitors shall be assigned to personnel who have disabilities so as to assist them in the evacuation of the premises. During a fire emergency, the use of elevators is forbidden unless assisted and manned by a fire department officer. Monitors shall be guided by the *special instructions for persons with disabilities* (assignment and details should be made part of this directive).

Stairway Monitor

Stairway monitors shall take their positions at an assigned stairway and calmly assist in the evacuation of personnel during an evacuation. In addition, they shall

- Inspect the stairwell for smoke and heat before and after evacuation of personnel
- Instruct personnel to form a single line down the stairway and to keep to the right side of the stairs
- Determine that all searchers have cleared all personnel and visitors from the floor
- Make sure that personnel do not use elevators during an evacuation

A list of the assigned wardens, searchers, and monitors will be made part of the Fire Safety Plan, posted and maintained up to date.

A copy of the floor plan shall be posted throughout all the floors and building, noting evacuation routes and the locations of fire alarm pull stations, fire extinguishers, and fire hoses within the building. A copy shall also be attached to and made part of the Fire Safety Plan.

All wardens, searchers, and monitors must remember to act calmly and without panic or hysteria. Be direct and assertive in your actions and voice commands; do not scream. If you show panic or indecisiveness, it will become contagious.

In the Event of a Fire

1. In the event of a fire, the fire warden shall determine the location and the extent of the fire.

2. For communications to the fire command station, a fire warden station should be designated and located on each floor level, usually by a fire stairway. The fire alarm pull box should be located in this same area, along with a telephone for internal and external communication.
3. The fire warden shall ascertain that the fire alarm pull station has been activated so as to transmit the fire signal to the local fire department, and that the building fire safety director is notified as to the particulars.
4. The fire warden shall alert the deputy fire wardens and searchers if not already so advised.
5. To prevent panic, the wardens and searchers must act decisively.
6. Direct the evacuation of personnel in accordance with directions and guidelines.
7. The fire warden shall advise the building's fire safety director of the means and routes to be employed by the occupants for evacuation.
8. Verify that all spaces have been cleared and evacuated by the searchers.
9. Personnel must be evacuated to two or more floor levels below or if unable to do so, two or more floor levels above the fire floor to be considered generally adequate for safety.
10. Determine that fire stairways are smoke free and uncontaminated before use by evacuating personnel.
11. Unless unavoidable, do not use fire stairways used by firefighting officers.
12. Do not use an elevator for evacuation unless it is manned by a firefighter who gives permission to enter.
13. Make use of a bullhorn, air horn, whistle, or other device to make sure that all occupants of the floor are advised to the fire threat and evacuation.

Reporting the Fire

If a fire is discovered or reported, and unless the Fire Safety Plan indicates that all reports of a fire should be conveyed to the security desk or the fire command station, the following general procedures by the floor warden should be taken:

1. Call the fire department immediately.
2. Inform the fire department dispatcher that you are calling to report a fire and *give the dispatcher your correct building street address.* If reporting a fire to the fire department and the fire is located other than from where you are calling, advise the dispatcher of the correct address of the fire location.
3. Advise the dispatcher of
   - What is on fire
   - Location of the fire and on what floor
   - Your name
   - Your company name
   - Your company's floor number and your telephone number

4. Do not hang up until you give the dispatcher all of the above information and listen for any additional instructions from the dispatcher.
5. Activate the nearest fire alarm pull station (if not already sounded) at the location shown on the attached floor plan
6. Call the security desk or fire command station at the correct extension number and advise of actions taken.

The Evacuation

Following the above initial report:

1. A fire alarm signal (bell, horn, etc.) will be activated for the whole building.
2. Begin to evacuate the floor and premises as dictated in the Fire Safety Plan.
3. The fire command station will contact the evacuation brigade, the fire floor warden, and floor wardens directly above and below the fire floor.
4. The floor wardens will advise the tenant warden, searchers, and monitors assigned to their floor, and shall assume full control of their floor.
5. The tenant wardens, searchers, and monitors shall go into immediate action to fulfill their assigned duties.
6. The evacuation brigade shall immediately occupy their stations and perform their assigned duties.
7. As soon as the occupants of the fire floor have reached a safe level of descent in the stairways, stairway monitors shall signal personnel in the adjacent floors immediately above and below the fire floor to evacuate. If occupants located in floors above the fire are unable to evacuate through the fire floor, the safest action is generally considered to be evacuation at least two floors above the fire.
8. Stairway monitors shall close all fire and stairway doors after the last person exits the floor.
9. Unless the fire stairway contains smoke or fire, floor wardens shall distribute the flow of people evenly via all available stairway exits.
10. If the fire stairway does contain fire or smoke, an alternative stairway may be reached via cross-over through the closest tenant floor.
11. All elevators will be called to the ground floor and put under manual control.
12. Any lobby congestion will be cleared by the evacuation brigade.
13. All personnel, once evacuated safely from the building, shall remain at least 300 feet from the building or as directed.

14. The alarm signal will continue throughout the state of emergency and will be silenced when the "all clear" is announced.
15. When the "all clear" is announced, members of the evacuation brigade will advise the floor wardens to lead the employees in an orderly fashion back to their workplaces.

General Safety Instructions

The following guidelines are to be followed by all employees:

1. When the fire alarm signal is sounded, a complete evacuation of the building will take place.
2. It will be the responsibility of the wardens, searchers, and monitors to include the control and direction of visitors, customers, etc., in evacuation procedures.
3. During an evacuation, the emergency (fire) floor shall be evacuated first and immediately. Following that, the next floors to be evacuated shall be the floor above and the floor below the emergency floor.
4. If notified of a fire on another level, never open any door to the fire floor.
5. Touch and feel all doors for heat before opening.
6. The fire safety director shall be in charge of the evacuation until the arrival of the fire chief.
7. Fire extinguishers are to be prominently located on each floor in the common corridor and at locations specified by the municipality's fire code. All tenant employees should familiarize themselves with the locations of all fire extinguishers on their floor.
8. All tenant employees should determine and familiarize themselves as to where the fire alarm pull stations are located on their floor.
9. All employees shall also familiarize themselves with the closest fire exit and fire evacuation routes in the vicinity of their offices.
10. Employees shall advise the fire warden of any defective, damaged, or missing fire extinguishers or any ineffective fire hardware as soon as observed. The fire warden shall advise the building's fire safety director immediately after being so advised.
11. All employees, and in particular the fire wardens, shall routinely determine that all doors to stairwells and fire stairways are self-closing and unobstructed, all stairways are unobstructed, all exit routes are kept clear and unobstructed, all directional fire exit lighting is operational, and that all emergency lighting and/or emergency generators are operational.

Weapons of Mass Destruction: Atomic, Biological, and Chemical Warfare

The following narrative is a brief description of the nature and effects of nuclear, biological, and chemical agents upon people, places and things.*

Introduction

Consider that medical defense against nuclear and radiological warfare is one of the least emphasized segments of modern medical education. In all the years since the first atomic device was used and the doomsday predictions that followed, management of casualties has not included any realistic preparations, adaptations, or procedures, mostly because of apathy and political considerations. Soon after World War II, there was some semblance of civil defense against an A-bomb attack via bomb shelters, both public and personal. School children during that period were taught to hide under their desks when and if an attack did occur. However, since the end of the Cold War, the mindset of the populace in America has dramatically reduced any worry about a thermonuclear war.

Unfortunately, the threat continues. Sometime before and since the fall of the Soviet Union, there has been a proliferation of atomic weapons in significant sovereign states and Third World nations. Particular attention must be given to these Third World countries, mainly because of fanatical, religious, and extremist views, and their will to annihilate anyone not of their ideology, particularly those Western nations who they believe have corrupted and interfered with their way of life.

The availability of other weapons of mass destruction, as described elsewhere in this book (see Chapter 7, "Weapons of Mass Destruction of the Second Class"), must be considered. Unless purchased from rogue nations or having the wherewithal and elements to manufacture a nuclear weapon, the only other means of inflicting a massive attack on the general public would be the use of biological or chemical weapons, which are more easily produced and dispersed.

Following is a brief description of the various properties of nuclear, biological, and chemical substances that are known and obtainable under various circumstances by those of criminal intent, depraved individuals, or the terrorists. Depending on the seriousness of the event and the amount of exposure disseminated all agents in the list below can cause severe illnesses, injuries, and death to large groups of people.

- Nuclear agents and substances

* The narrative has been extracted from the Department of Homeland Security, Centers for Disease Control, and Health and Human Services Web sites and from *The United States Armed Forces Nuclear, Biological and Chemical Survival Manual*, by Dick Couch, Captain, USNR (retired), ©2003, published by Basic Books, New York, New York.

- Biological agents and substances
- Viral agents
- Biological toxins
- Chemical agents and substances

Agents and Elements Defined

The purpose of an atomic attack is to cause massive damage, death, and injury by blast and thermal energy over a large geographic area, along with radioactive exposure covering an even larger area so as to cause acute and chronic radiation sickness and death. Aside from the initial and inevitable devastation following a nuclear assault, the populace following such an event would suffer, initially at least, panic, fear, apprehension, depression, and a dispirited disposition to rise up and fight back.

A nuclear bomb may not be the only dispersal instrument for radiation exposure. As discussed in Chapter 7, a radiation dispersal device (RDD) (more commonly known as a "dirty bomb") would contain radioactive material that may be scattered by a simple conventional explosive, thereby causing serious radioactive harm. Though the area of contamination will not be as great as that which would have been spread by an atomic bomb, the seriousness of an incident of this type cannot be overstated. Authorities experienced in terrorist activity believe that a terrorist employing the use of a dirty bomb (a nonfission device) containing plutonium and uranium can suffer little risk of exposure or harm to himself, have easier access in obtaining radioactive nuclear waste, and would need less expertise and technology to build such a device. Therefore, a mechanism of this type would be easier to manufacture than an atomic bomb, which requires sophisticated expertise and materials.

An electromagnetic pulse bomb (also covered in Chapter 7), could wreak severe and lasting damage to electric and electronic devices. This would include computers, communications (radio, telephone, and satellite), transportation systems, and financial services, to name a few. Moreover, a crippling effect to communications would also cause medical, emergency, and military responses to be delayed to a great degree. An atomic bomb would cause similar havoc on electronic devices, but the "pulse bomb" is deployed specifically for that particular purpose. The reasoning is that the capture of an area without the use of nuclear devices, with their attendant atomic-produced radioactive contamination, would be more advantageous for an occupational force.

Other Sources of Radioactive Contamination

Consider also that radiological contamination can occur by other means. Conditions could include sabotage or accidental release of radiation from radiological sources such as nuclear waste processors, nuclear power plants, research and medical facilities, industrial complexes, and pirated weapons-grade nuclear material (uranium and plutonium-239). The radiation can be scattered across the targeted area as an aerosol or particulate debris, depending on the weather conditions. The release of radiation in these scenarios would complicate medical evacuation and treatment of casualties, since one of the objectives of the terrorist is to function as a terror weapon and prevent habitation by contamination.

The Ultimate Threat

The illness, injury, death, and mass panic created by dispersal of radiation in a populated area will occur through inhalation, ingestion, or skin exposure. Consider also the effect of radiation material deposited on exposed surfaces, plants and foodstuffs.

Other than the nuclear blast that will cause immediate damage, death, and injury within a certain geographical area, the more long-lasting illness and death will occur because of *ionizing radiation exposure* and *radioactive contamination*. This may be considered of greater concern because of the larger area and long-lasting contamination effect on the civilian population exposed to radiation levels that are harmful.

Radiation, particularly gamma rays, damages cells in living tissue through ionization, destroying or altering the cell elements essential to a living organism. Depending on the level of radiation, exposure may be acute (immediate or within a short period) or chronic (over a long or drawn-out period), and people will become sick or die.

Therefore, we must consider that significant amounts of radioactive material will be deposited on exposed surfaces after the use of a nuclear weapon or RDD, destruction of a nuclear reactor, a nuclear accident or improper nuclear waste disposal.

Types of Ionizing Radiation

Other than people close enough to a nuclear detonation who would die due to blast and thermal effects, the following radioactive elements will cause illness or death in various degrees:

Alpha particles: These particles travel only a few centimeters and present hazard when inhaled or ingested. Because of their size, massive charged particles (four times the size of a neutron) cannot travel far and can be fully stopped by dead layers of skin or clothing. These particles can be negligible unless injected or inhaled into the lungs and gastrointestinal tract, which will be a cause of significant cellular damage in that area.

Beta particles: These particles travel a few meters in the air and have limited penetration power, and therefore present danger to the skin and eyes. These charged particles are very light and are primarily found in fallout radiation.

They can travel a short distance in tissue, and if large quantities are involved, they can produce severe damage to skin tissue by producing lesions (known as "beta burns") similar to a thermal burn. It appears that the most sensitive organ to beta particles is the lens of the eye.

Gamma rays: These particles travel at the speed of light and can only be shielded by heavy materials such as lead, steel, and concrete. These rays are emitted during a nuclear detonation and in fallout. They are uncharged radiation particles similar to x-rays, highly energetic, and pass through matter easily. Because of its high penetrability, gamma radiation can result in whole-body intense exposure.

Neutrons: These can travel anywhere from hundreds of yards to several miles, and consist of heavy neutral particles acting as millions of tiny bullets penetrating the human body and causing severe cell damage. Like gamma rays, they are uncharged, and they are only emitted during a nuclear detonation. However, they are not fallout hazards, but have significant mass and interact with the nuclei of atoms, thereby severely disrupting atomic (microscopic) structures. They can cause 20 times more damage to tissue than gamma rays. Neutron radiation, although potentially present in very minute amounts of uranium and plutonium, is not likely to be found in a "dirty bomb."

Radiological Effects

When radiation interacts with atoms, energy is deposited, resulting in ionization (electron excitation). This action may damage certain critical molecules or structures in a cell (human, animal, or plant). The two most significant entities in the human body sensitive to radiation are the hemoglobin (blood) and the gastrointestinal system. However, radiation may attack particularly sensitive atoms or molecules in any cell. Damage to a cell is irreversible; the cell will die or malfunction. Other than the immediate blast and thermal injury, clinical symptoms of radiation will occur in the major systems such as bone marrow, gastrointestinal, and neurovascular. Radiation sickness and/or death can occur within 24 hours or 6 or 8 weeks after exposure, depending on the amount and type of radiation and the subject's resistance to the dose received. The presence of other injuries will increase the severity and accelerate the infection suffered.

Treatment

Medical attention for blast, thermal, or radiation injuries must be as immediate as possible, realizing of course that medical and emergency responders will also be exposed to radiation unless protective gear is worn and other precautions are met. These and other conditions will slow the response time in aiding those who are injured.

Management of injuries will be divided into three phases:

Triage: Patients are prioritized and rendered immediate lifesaving care.
Emergency care: Therapeutics and diagnostics are necessary during the first 12 to 24 hours.
Definitive care: Final disposition and therapeutic regimens are established.

The primary goal is to evacuate radiation casualties from contaminated areas prior to the onset of evident illness. Treatment will depend on the severity and number of injuries sustained by the populace, the available facilities, the number of medical personnel, and available medical resources. Management and conventional therapy of radiation sickness include the prevention and control of infection with the use of antibiotic prophylaxis to reduce pathogen (microorganism) acquisition and, therefore, infection. All types of microbial invasions will be considered as pathogens, and susceptibility varies with time.

Biological Agents and Substances

The characteristics of biological weapons as used in warfare are as follows:

- A potential for massive numbers of casualties
- Ability to produce lengthy illnesses that require long-term and extensive care
- Contagion
- Lack of adequate detection systems
- Diminished available aid, thereby increasing the sense of helplessness
- An incubation period enabling victims to disperse widely and infect others
- Ability to produce nonspecific symptoms or to mimic other infectious diseases, thereby complicating diagnosis

Bacterial Substances

Bacteria cause disease in humans and animals by invading the host tissues and by producing poisons (toxins). Many pathogenic bacteria utilize both hosts to cause harm. Diseases produced in this manner often respond to specific therapy with antibiotics. The following are to be considered potential biological warfare threat agents:

Anthrax: Anthrax is primarily a disease of herbivores: cattle, sheep, goats, and horses are the usual domesticated animal hosts. Although infection may be introduced through scratches, abrasions, and open wounds on the skin, the primary concern for intentional infection by the anthrax organism is through aerosol dissemination of spores. These spores are easily cultivated;

are very stable and highly resistant to sunlight, heat, and disinfectants, and may remain viable for many years in soil and water. This agent can be disseminated by missile attacks, high-flying aircraft, or any spray or aerosol device. Recent dissemination had a beginning on September 18, 2001 and continued for several weeks thereafter. Letters were sent via the U. S. Postal Service containing anthrax powder (5 cases) and received at the offices of several news media and two U. S. Senators. It culminated in 23 confirmed cases of inhalational anthrax, which included 5 deaths and 18 ascertained sicknesses.

A suspect was investigated and finally named as the sole culprit by the F.B.I in 2008. The suspect, Bruce Edwards Ivins, a scientist employed at the Government's Biodefense Labs at Fort Detrick, Detrick, Maryland, committed suicide on August 1, 2008.

Symptoms and treatment: incubation period generally 1 to 6 days. Fever, malaise, fatigue, cough, and chest discomfort, progressing to severe respiratory distress: difficult or labored respiration, profuse perspiration, a harsh vibrating sound heard during expiration in cases of obstruction of the air passage, a blue or purple color to the skin due to deficient oxygenation of the blood, and shock. Although treatment may be limited after symptoms are present, high-dose antibiotics such as penicillin and oral ciprofloxacin (Cipro) are used to treat for known or imminent exposure. At this time, Cipro is considered the first line of defense. However, after significant exposure and symptoms have advanced, inhalation cases have proved fatal regardless of treatment. Terrorist experts believe that an adversary can produce anthrax with a resistance to ciprofloxacin, penicillin, and other antibiotics through laboratory manipulation of the spores.

Brucellosis: This illness is typically observed as a fever, headache, back pain, sweats, chills, depression, mental aberrations, and general uneasiness. It is uncommon for fatalities to occur. First observed in British soldiers on Malta during the Crimean War, the *B. melitensis* organism produced Mediterranean gastric remittent fever. Goats were identified as the source through ingestion of unpasteurized milk products. Although animal vaccines exist, there is no human vaccine. Transmission between humans can occur via sexual contact. Human brucellosis is an uncommon disease in the United States. Most human cases are caused by ingestion of unpasteurized dairy products and affect those employed in the veterinary field. At this time, antibiotic therapy in combination with other medications is the treatment of choice. Considered one of the most important veterinary diseases of the reproductive tract, brucellosis causes septic abortion and sterility. Consequently, this disease has a great potential economic impact to animal husbandry.

Treatment: There is no vaccine available for human use. Depending on the severity of the infection, various antibiotic therapies are indicated for periods up to 6 months.

Plague: Pneumonic plague begins after an incubation period of 1 to 6 days with high fever, chills, headache, and general malaise, followed by cough, progressing rapidly to death. Gastrointestinal symptoms are often present. Death results from respiratory failure, circulatory collapse, and internal bleeding. Plague is highly contagious. ***Bubonic plague*** features high fever, malaise, and painful inflammatory swelling of the lymph nodes, particularly in the area of the groin. Infection may progress spontaneously to septic shock and/or thrombosis. This disease has been transmitted by fleas living on rodents (rats, mice, squirrels), which then transmit the bacteria to humans, who then suffer the bubonic form of the disease which may progress to pneumonic plague. Transmission can be obtained also by human-to-human contact, particularly via respiratory droplets. Weaponization of the pneumonic form is effective via aerosol dissemination and exposure. No vaccine is presently available. An early vaccine was effective against bubonic plague, but not against aerosol exposure.

Treatment: Early administration of antibiotics is critical, as pneumonic plague is fatal if antibiotic therapy is delayed more than one day after the onset of symptoms.

Q fever: Infection is not a clinically distinct illness and may resemble a viral illness such as pneumonia. A vaccine is effective against exposure, but severe local reactions to this vaccine will occur in those already possessing immunity. Therefore, a skin test is recommended to detect presensitized or immune individuals. Natural reservoirs of this disease are sheep, cattle, goats, dogs, cats, and birds. These animals do not develop this disease but do shed the organism, as it grows in high concentrations in placenta tissues, body fluids, milk, urine, and feces. Humans acquire this disease by inhalation of aerosols contaminated with the organisms. A biological warfare attack with Q fever would cause a disease similar to that occurring naturally, and, therefore, an adversary would make use of this disease as an incapacitating biological warfare agent.

Treatment: Most cases of acute Q fever will resolve without antibiotic treatment. Vaccine is available for immunization of at-risk medical and emergency personnel. However, as noted above, administration of this vaccine to immune individuals may cause severe reactions.

Tularemia: Also known as rabbit fever and deerfly fever. This disease is usually acquired by humans via contact with tissues or body fluids of infected animals or, more commonly, by bites of infected ticks, deerflies, or mosquitoes. Less common is by inhalation of contaminated dusts or ingestion of contaminated foods or water. Respiratory exposure by aerosol would typically cause ***typhoidal*** or ***pneumonic tularemia***. Weaponization of this disease has been known to be accomplished by other countries. ***Ulceroglandular tularemia*** is observed with ulcers and inflammatory lymph glands, fever, chills, headache, and malaise. ***Typhoidal tularemia***

is noted by fever, headache, malaise, discomfort, exhaustion, weight loss, and a persistent cough.

Treatment: Appropriate antibiotic therapy is indicated, with strict adherence to standard precautions for draining lesions and disinfection of soiled clothing, bedding, and equipment.

Viral Agents

Viruses are the simplest microorganisms, much smaller than bacteria. Viruses are intracellular parasites, are dependent on the host cells they attack, and therefore lack a metabolism. This means that viruses, unlike bacteria, cannot be cultivated, but require living cells from a host to multiply. Host cells may be from humans, animals, plants, and even bacteria. Typically, a virus brings about changes in the host cell that eventually lead to cell death and, ultimately, death to the host.

There are three types of viruses that can be employed in biological warfare:

1. Smallpox
2. Alpha viruses (Venezuelan equine encephalitis, for example)
3. Hemorrhagic fever viruses

Smallpox

Endemic smallpox had been declared eradicated in 1980 by the World Health Organization (WHO). Although there are two repositories authorized by WHO (Centers for Disease Control and Prevention in Atlanta and the Institute for Viral Preparations in Moscow), whether clandestine stockpiles exist elsewhere in the world is unknown. However, despite the eradication of the smallpox virus, in the opinion of disaster authorities, smallpox would be the ***virus of choice for a terrorist group***. The virus occurs in at least two known strains: ***variola major*** and the milder form, ***variola minor***.

Symptoms begin acutely with malaise, fever, rigors, nausea, vomiting, headache, and backache. Within two to three days, lesions appear, which will progress from macules (raised yellowish cystlike blemishes) to papules (small conical elevations of the skin), and eventually to pustular vesicles (elevated cystlike or blisterlike papules containing pus and an inflamed base). These lesions appear more abundantly on the extremities and the face, and are characteristic of this infection. Droplet and airborne precautions must be placed in effect. Nevertheless, close person-to-person contact is the general rule for transmission of this disease. Any confirmed case of smallpox should be considered an international emergency, with an immediate report made to the proper public heath authorities.

Treatment: Strict quarantine of those infected, along with immediate vaccination or revaccination of the smallpox vaccine. Isolation is required until all scabs separate from the body. Smallpox vaccination (*vaccinia virus*) with a verified positive effect (vesicle with a scar formation on the skin) within three years is considered to render a person immune to smallpox.

Venezuelan Equine Encephalitis

This virus complex is caused by mosquito-borne alpha viruses endemic to northern South America and Trinidad and is also the cause of rare cases of encephalitis. These viruses can cause severe diseases in humans and animals (horses, mules, burros, and donkeys). There is no evidence of human-to-human or horse-to-human transmission.

Those infected suffer malaise, spiking fevers, rigors, headache, photophobia, and myalgias for approximately 24 to 72 hours. Nausea, vomiting, cough, sore throat, and diarrhea may follow. Full recovery will take approximately 1 to 2 weeks. Acute systemic fever with encephalitis can develop in a small percentage of children (4%) and less than 1% in adults. Patient quarantine is not required. The incidence of this disease causing death would be much higher in the case of a biological attack.

Treatment: Uncomplicated infections may be treated with analgesics. Patients who develop encephalitis may require anticonvulsants and intensive supportive care to maintain fluid and electrolyte balance, and to avoid any complicating bacterial infection. At present there are two investigative vaccines in development and research continues.

Viral Hemorrhagic Fevers

VHFs are feverish illnesses that feature flushing of the face and chest, minute hemorrhage of the skin, bleeding, edema, hypotension, and shock. As with all hemorrhagic fevers, malaise, myalgias, headache, vomiting, and diarrhea may occur. The viral hemorrhagic fevers are a diverse group of illnesses caused by four viral families:

Arenaviridae: include the etiologic agents of Argentine, Bolivian, and Venezuelan hemorrhagic fevers, and Lassa fever transmitted to humans by the inhalation of dusts contaminated with rodent excreta.

Bunyaviridae: include the Hantavirus (rodent-borne and inhalation of dusts contaminated with rodent excreta), the Congo-Crimean hemorrhagic fever from the Nairovirus (spread by infected animals and health-care settings), and the Rift Valley fever from the Phlebovirus (mosquito-borne)

Filoviridae: include the Ebola virus (found to be spread from monkey to monkey, and monkey to human via the respiratory route) and the Marburg virus (apparently spread by direct contact with infected blood, secretions, organs, or semen).

Flaviviridae: include dengue and yellow fever viruses (both mosquito-borne).

These viruses are spread in a variety of ways. Some may be transmitted to humans via a respiratory portal of entry. Although it is unknown whether these viruses can be utilized in biological warfare, they present potential for weaponization in aerosol dissemination.

Treatment: The only licensed VHF vaccine is yellow fever vaccine. No specific viral therapy exists. Antiviral therapy with ribavirin may be effective for the other viruses in this category. Uncomplicated exposure may require only analgesics. Also, in some of these viruses, strict isolation with appropriate sterile garments and respiratory and eye protection of medical personnel coming within 3 feet of the patient is indicated.

Biological Toxins

Toxins are harmful substances produced by living organisms such as animals, plants, and microbes. They are distinguished from chemical agents such as cyanide or mustard, which are human-made and nonvolatile (no vapor hazard), usually not noxious to skin exposure, and generally much more toxic per weight than chemical agents. Because toxins lack volatility, they are unlikely to produce either secondary or person-to-person exposures or any persistent environmental hazard.

The major effect of a toxin as an aerosol weapon is determined by its toxicity, stability, and ease of production. The bacterial toxins, such as *botulinum toxins* are the most toxic substances by weight known. Less toxic compounds are many times, sometimes a thousand times, less toxic than botulinum. In some cases, stability limits the open-air potential of some toxins as a weapon. As with all biological weapons, the potential to cause incapacitation as well as death must be considered. Incapacitating agents may be more effective than lethal agents due to the overwhelming demand for medical and evacuation assistance, and the subsequent panic of the exposed population.

The following four toxins are considered to be the most likely to be used against military and civilian targets by a terrorist:

- Botulinum
- Ricin
- Staphylococcal enterotoxin B
- Mycotoxins

Botulinum

The botulinum toxins are seven related neurotoxins produced by the spore-forming bacillus ***Clostridium botulinum*** and two other ***Clostridia*** species. These toxins, classified as types A through G, are the most potent neurotoxins known. Botulinum can be delivered by aerosol or used to contaminate food and water supplies. The clinical syndrome produced by these toxins is known as *botulism*. According to national data, several countries and terrorist groups have been known to produce these toxins as an aerosol biological weapon. Symptoms include respiratory failure due to paralysis of respiratory muscles—the most serious effect, and generally the cause of death.

Treatment: Prior to 1950, botulism had a mortality rate of 60%. With prompt respiratory support such as tracheotomy, intubation, and ventilatory assistance, fatalities are less than 5% today. Symptoms that continue to progress will be helped by an early administration of botulinum antitoxin, which is critical. After symptom progression ceases, the antitoxin will have no effect. Postexposure treatment for animals is the use of heptavalent antitoxin, but no human data is available at this time. Intensive and prolonged nursing care may take up to three months for initial signs of improvement, and up to a year for a complete recovery.

Ricin

Symptoms include the onset of a fever, tightness of the chest, cough, nausea, and labored respiration within 4 to 8 hours of inhalation exposure. Following that, pulmonary edema will likely occur, causing severe respiratory distress and death within 36 to 72 hours. When ingested, ricin will cause severe gastrointestinal symptoms followed by vascular collapse and death. However, aerosol use appears to be the weapon of choice. Use of a protective mask is currently the best protection against the inhalation of this agent. Because of its ready availability, relative ease of extraction, and past notoriety in the press, the potential of use by a terrorist group is feasible.

Treatment: Includes management of pulmonary edema. There is no current vaccine or antitoxin available for human use, although immunization appears promising in animal models.

Staphylococcal Enterotoxin B

SEB is one of the toxins that commonly cause food poisoning in humans after the toxin is produced from improperly handled foodstuffs that are ingested. After aerosol exposure, and a latent period of about 3 to 12 hours, there is a sudden onset of fever, chills, headache, and myalgia; if ingested, nausea, vomiting, and diarrhea will follow. A different clinical syndrome is caused when the toxin is ingested or inhaled. Some patients will develop shortness of breath and chest pain. If the exposure has been high, septic shock and death will occur. Although an aerosol SEB toxin weapon would not likely produce a significant death rate, it could render 80% of the exposed populace clinically ill and unable to perform routine activities for up to two weeks. Therefore, the demand for medical and logistical support would be overwhelming.

Treatment: There is no human vaccine for immunization against SEB intoxication, although several candidates for a viable vaccine are in development. Close attention to pulmonary edema is indicated. Fever and cough suppressants may make the patient more comfortable.

Mycotoxins

This agent is generally used as an aerosol in the form of "yellow rain," with droplets of variously pigmented oily fluids contaminating clothing and the environment. Yellow rain has been reported to have been released from aircraft in Laos (1975–1981), Kampuchea (1979–1981), and Afghanistan (1979–1981). Total deaths in these three incidents were approximately 10,500.

Exposures by mycotoxins cause dermal, ocular, respiratory, and gastrointestinal effects. Exposure causes skin pain, severe itching of the skin, fluid-filled blisters, death of localized skin tissue, and the shedding of dead skin from the body. Airway infection includes nose and throat pain, cough, wheezing, and chest pain, with severe intoxication resulting in weakness, collapse, shock, and death.

Treatment: There is no specific antidote. Washing the infected skin with soap and water can reduce dermal toxicity significantly. If the toxin has been swallowed, activated charcoal should be given orally. Outer clothing that has been exposed should be removed immediately, and exposed skin cleansed with soap and water. Eye exposure requires copious saline irrigation.

Chemical Agents and Substances

Pulmonary Agents

Phosgene: First used as a chemical warfare agent on the battlefield in 1917 during World War I. Employed initially by Germany by filling shells solely with phosgene, later both sides employed phosgene alone or as a mixed-substance shell, usually with chlorine. Although this chemical was manufactured and stockpiled by the U.S. military in World War II, it was never used. The presence of this agent, via a vapor or a liquid, can be detected by its odor: newly mown hay or freshly cut grass. Contamination is by inhalation. Symptoms include eye and airway irritation, difficulty breathing and tightness of the chest, and pulmonary edema (which may become chronic).

Treatment: Immediate termination of the exposure followed by decontamination depending on the state and type of exposure: if vapor, fresh air; if liquid, copious water irrigation followed by resuscitation protocol. Management of the patient's airway secretions and lung congestion is warranted. The patient must be forced to rest; even minimal physical exertion will increase the severity of the respiratory symptoms, which will cause acute deterioration and possibly death.

Blood Agents

Cyanide and chloride: Cyanide is a fast-acting lethal agent that is limited for military use because of its high volatility. Categorized as a blood agent, the designation appears to be a misnomer. Other chemical agents in use caused mainly local effects such as injuries to skin and mucous membranes from direct contact, in addition to damage to the lungs after inhalation of phosgene. In contrast, inhaled cyanide produces systemic effects and was thought to be carried in the blood; hence the term *blood agent*, still used today. It was introduced by the French during World War I without success and poor results. The high volatility of cyanide causes quick evaporation and dispersal, so that extra-large amounts of the agent are needed to cause any biological effect. Exposure is by a vapor in the air. Skin exposure is usually not that serious because the agent is so volatile. However, wet contaminated clothing should be removed and the underlying skin decontaminated with water and other standard decontaminant. After exposure to high-enough doses, symptoms include seizures, followed by respiratory and cardiac arrest. Death will occur within 6 to 8 minutes after inhalation of a high dose.

Treatment: Immediate removal of patient to fresh air. Skin decontamination is not necessary unless as noted above. Basic medical management is by intravenous sodium nitrate and sodium thiosulfate. Administration of pure oxygen appears to help.

Vesicants

Mustard: Sulfur mustard, first used in World War I and commonly known as mustard gas, caused serious injury to the combatants who were exposed. Mustard produced most of the chemical casualties in 1917 by Germany. Effects of exposure may be delayed, appearing hours later, with symptoms including skin irritation and blisters, conjunctivitis, severe injury to the eyes, mild upper respiratory signs to severe airway damage, along with gastrointestinal effects and bone-marrow cell suppression. Agents of this type pose a specific threat via vapor and liquid to all exposed skin and mucous membranes.

Treatment: There is no specific antidote. Immediate decontamination is the only way to reduce physical damage. Skin treatment by lotions will reduce burning and itching. Treatment for eye exposure should begin with complete irrigation of the eyes, along with ophthalmic solutions and ointments. Pulmonary symptoms may respond to steam inhalation and cough suppressants.

Lewisite: A vesicant that damages the eyes, skin, and airways by direct contact. Absorption causes immediate pain or irritation, shock, and organ damage. Lewisite is sometimes mixed with sulfur mustard to achieve a lower freezing point of the mixture for ground dispersal and aerial spraying. Airway damage and skin blisters are similar to sulfur mustard exposure.

Treatment: Immediate decontamination is necessary to prevent or lessen damage. This immediate action will require self-aid rather than later medical management. Skin lesions may be treated as with exposure to sulfur mustard. A specific antidote (British anti-lewisite [BAL]) is available but may cause some toxicity, and the user may wish to read the package inset carefully before using.

Phosgene oxime: This agent causes a corrosive type of skin and tissue lesion. It is not considered to be a true vesicant, since it does not cause blisters. The vapor is extremely irritating, and both the vapor and liquid cause almost immediate tissue damage on contact. Little else is known of this agent, and there is no current assessment of a potential military threat.

Treatment: Immediate decontamination with water and symptomatic management of lesions.

Nerve Agents (Tabun, Sarin, Soman, GF, and VX)

Nerve agents are the most toxic of known chemical agents. They are extreme hazards in a liquid and vapor state, can produce rapid symptoms, and will cause death within minutes of exposure. Nerve agents can be dispersed by artillery or mortar shells, rockets, land mines, missiles, aircraft sprays, or aircraft bombs or bomblets. They can also be hand-carried by a terrorist in a can or as an aerosol.

Symptoms include the following:

Vapor

Small exposure: mild difficulty breathing, eye impairment, inflammation of the mucous membrane of the nose

Large exposure: sudden loss of consciousness, convulsions, apnea, inflammation of the mucous membrane of the nose, paralysis, and copious secretions

Liquid on skin

Small to moderate exposure: localized sweating, nausea, vomiting, feeling of weakness

Large exposure: sudden loss of consciousness, convulsions, apnea, paralysis, copious secretions

Treatment: decontamination with hypochlorite and large amounts of water, ventilation, administration of known antidotes, and supportive therapy.

Incapacitating Agents (Bz, Agent 15)

These agents cause excessive dilation of the pupil of the eyes, dry mouth, dry skin, decreased level of consciousness, confusion, disorientation, perception and interpretation is diminished, short attention span, and impaired memory. Dispersal can be by aerosol for inhalation or dissolved in food or liquid for ingestion. The greatest risks to life may arise from erratic demeanor or hyperthermia in patients who are located in hot, humid environs and dehydrated from overexertion or insufficient water intake.

Treatment: Removal of clothing, along with a thorough flushing of skin and hair with soap and water, is required. The antidote of choice is physostigmine.

Summary

The Department of Homeland Security is the lead federal agency responsible for responding to the full range of terrorist threats and the unification of our defenses against human, animal, and plant diseases that can be used as terrorist weapons. The department will also set national policy and establish guidelines for state and local governments. It will be the leader in establishing research, development, and testing to invent new vaccines, antidotes, diagnostics, and therapies against biological and chemical agents. Moreover, the department must recognize, identify, and confirm the occurrence of an attack and minimize the morbidity and mortality caused by any nuclear, biological, or chemical agent.

Investigation, intelligence, and research and development by all subagencies controlled by the Department of Homeland Security are driven by a constant examination of the nation's vulnerabilities, constant testing of our security systems, and a constant evaluation of all threats and weaknesses based on is the probability of a catastrophic terrorist event that could result in a large-scale loss of life, incapacitating illnesses, a disruption of our way of life, and a major economic impact.

However, since some operations of Homeland Security leave much to be desired at this time, it is expected that continuous oversight and evaluation of vulnerabilities and threats and interconnection of all defenses will be considered the standard to follow.

Coastal Areas of the United States

Geographical Regions Considered Hazardous to Life and Property and Conducive to Disastrous Weather Conditions

As defined by the National Oceanic and Atmospheric Administration (NOAA), a county is considered a coastal watershed if one of the following criteria is met:

1. At a minimum, 15% of the county's total land area is located within a coastal watershed
2. A portion of the entire county accounts for at least 15% of a coastal cataloging unit (i.e., an individual drainage basin)

Additionally, a coastline is defined as any land that borders the ocean and any of its saltwater tributaries, including bays and tidal rivers, and the Great Lakes.

As of Census 2000, the following numbers of people and housing units are within 50 miles of a coastline:

- 137.5 million people (48.9% of the U.S. total)
- 55.4 million housing units (47.8% of the U.S. total)
- 1.7 million seasonal housing units (48.1% of the U.S. total)

Since 1950, the coastal population has grown from 75.2 million to 155.2 million in 2005. The overall population increase between 1950 and 2005 was 106.1% for coastal areas and 75.8% for noncoastal areas. Based upon census figures from 1950 to 2005, the number of Americans residing in coastal counties passed the 150 million mark, making the coastal population larger than the entire U.S. population in 1950. In 2005, there were 673 coastal counties in the 50 states and the District of Columbia, grouped into five regions:

1. The North Atlantic
2. The South Atlantic
3. The Gulf of Mexico
4. The Pacific
5. The Great Lakes

Today, NOAA calculates that more than half of the U.S. population lives in a coastal area as defined, even though the 673 coastal counties constitute only about one-quarter of the country's land mass.

In many cases, coastal counties have grown rapidly in otherwise slow-growing states. As an example, the United States Census Bureau notes that the population of Massachusetts increased by 11.6% between 1970 and 2000, while three of its coastal counties more than doubled in population during that same period.

Florida and Alaska were the only two states whose coastal county populations more than doubled—up 135.8% and 111.8%, respectively, between 1970 and 2000. In that same period, 134 coastal counties more than doubled in population, while 17 coastal counties in Florida had a population increase of more than 250%. Numerically, coastal counties gained 38 million people between 1970 and 2000, with 21.8 million of the increase in the Atlantic coastal region, 14.9 million in the Pacific coastal region, and 1.3 million in the Great Lakes coastal region.

The growth in population of coastal areas illustrates the importance of emergency planning and preparedness for areas susceptible to severe weather conditions. This is particularly so because of increasing concerns about global warming and the increase of abnormal weather conditions.

Source: The above narrative was compiled from the National Oceanic and Atmospheric Administration and the United States Census Bureau.

Volcano and Earthquake History*

Volcanoes

Few people think about the possibility of a volcano erupting; however, when one does, it becomes a major event. Of the 50 states in our union, 13 have volcanoes. Many are extinct or inactive, but there are enough to be of some concern, particularly during earthquake activity.

Alaska, 96	California, 26	New Mexico, 21	Arizona, 16	Hawaii, 16
Oregon, 43	Washington, 12	Idaho, 7	Utah, 6	Wyoming, 3
Nevada, 2	Colorado, 1	South Dakota, 1		

All known volcanoes are located in the western half of the United States. The most active volcano is Kilauea on the big island of Hawaii, which is continuously erupting lava flows as of this date. Many of those noted above are closely watched and monitored by the U.S. Geological Survey, an agency of the U.S. Department of the Interior.

The most recent eruptions (since 1990) and their locations are noted:

Alaska:	Mt. Akutan, 1992	Mt. Amukta, 1997	Mt. Augustine, 2006	Bogoslof Island, 1992
	Mt. Chiginagak, 1998	Mt. Cleveland, 2001	Mt. Kanaga, 1995	Mt. Kiska, 1990
	Korovin volcano, 1998	Mt. Makushin, 1995	Mt. Okmok, 1997	Mt. Pavlof, 1997
	Mt. Redoubt, 1990	Mt. Seguam, 1993	Mt. Shishaldin, 2004	Mt. Spurr, 1992
	Mt. Veniaminof, 2004	Mt. Westdahl, 1992	Fourpeaked, 2007	

* From Wikipedia, The Free Encyclopedia, "The List of Volcanoes and Earthquakes in the United States of America," accessed September 24, 2007; http://en.wikipedia.org.

Hawaii: Loihi Seamount, 1996 Kilauea, continuous eruptions

Washington: Mt. St. Helens, 2005, 2006, and June 2007; a new lava dome continues to enlarge, with continuous emissions of gas and ash

Super Volcanoes

Super volcano is a popular term for a large volcano that has the potential to cause severe cooling of global temperatures for many years because of the huge volumes of sulfur and ash erupted. An example is the Yellowstone Caldera, located in Yellowstone National Park in the western United States. It is currently inactive, but there are indications that of eruption may take place in the near future.

Decade Volcanoes

Sixteen volcanoes are currently designated as "decade volcanoes." Two volcanoes are worth noting because of their history of large destructive eruptions and their proximity to populated areas: Mauna Loa in Hawaii and Mt. Rainer in Washington.

Earthquakes

Earthquakes are much more common in our country than most people are aware, particularly along our West Coast and the Pacific Rim. Although there may be as many as hundreds per year, very few are severe enough to cause serious damage and loss of lives. Earthquakes occur almost every day throughout the world. The following list denotes the most recent earthquakes in the United States since 2000, as reported by the United States Geographical Survey (USGS) that monitors earthquakes worldwide.

Aug. 14, 2007	Hawaii	5.4 magnitude
July 20, 2007	San Francisco, CA	4.2 magnitude
Oct. 20 2006	Northern California	4.5 magnitude
Oct. 15, 2006	Hawaii	6.7 magnitude
Oct. 2, 2006	Maine	3.8 magnitude
Sept. 28, 2004	Parkfield, CA	6.0 magnitude
Dec. 22, 2003	San Simeon, CA	6.6 magnitude
Nov. 17, 2003	Rat Island, AK	7.8 magnitude
Nov. 3, 2002	Denali National Park, AK	7 magnitude
Apr. 20, 2002	Au Sable Forks, NY	5.2 magnitude
Feb. 28, 2001	Olympia, WA	6.8 magnitude
Sept. 3, 2000	Napa, CA	5.0 magnitude

Most earthquakes in the United States occur west of the Rocky Mountains. However, the most active areas east of the Rockies are the Eastern Tennessee Seismic Zone and the New Madrid Seismic Zone. On April 29, 2003, a 4.6-magnitude quake struck at Fort Payne, Alabama. It was felt in 11 states across the eastern United States. On December 9, 2003, a 4.5-magnitude quake struck an area near Richmond, Virginia. It was felt in five states and Washington, DC.

Competitive Intelligence: Company Profiles

Intelligence gathering under certain clandestine conditions may not be considered espionage. One area for in-depth research of businesses is the Internet. The Web makes a great place for competitive intelligence research. Many times, a search of a particular company will bring forth an exceptional amount of information, along with data that may have been inadvertently or unknowingly posted on the Web. Information that may also be found online would include adverse or harmful remarks by a disgruntled employee or some individual wishing to harm the subject company. A competitor searching for information under these conditions may have hit the jackpot for the information sought.

Looking for information on a business is as easy as inserting the company name (e.g., www.companyname.com). However, more pertinent information may be gathered through a variety of search engines and specific Internet business sources. The following resources are excellent sources of industry research:

Society of Competitive Intelligence Professionals

www.scip.org

Internet Intelligence Index

www.fuld.com/i3/index.html

Business and Economic Database (under Yahoo Companies, click on "Search Only")

www.yahoo.com/Business_and_Economies/Companies

Edgar (Electronic Data Gathering, Analysis and Retrieval System)

www.sec.gov

Hoover's Online (ultimate source via company name, keyword, or ticker symbol)

www.hoovers.com

Database America (research by name, geography, SIC code, individual contacts, number of employees, and sales volume)

www.databaseamerica.com

In addition, researching the electronic and print media for current and historical articles can reveal solid information, or at least lead an investigator to other sources. This is particularly true for local TV and newspapers, which offer regional insight.

News Wires

Business Wire	www.buinesswire.com
PR Newswire	www.prnewswire.com
Reuters	www.reuters.com

National News Sources

CNN	www.cnn.com
MSNBC	www.msnbc.com
USA TODAY	www.usatoday.com
The New York Times	www.nytimes.com
Los Angeles Times	www.latimes.com
Washington Post	www.washingtonpost.com

Primary Sources

Experts in the profession researched	
Conferences and event data	www.eventsource.com
Patents, IBM patents server	www.patents.ibm.com
U.S. Patent and Trademark Office (USPTO)	www.uspto.gov

Appendix B

Training in Security and Safety

The Loss-Prevention Manager's Role in Training

One of the most important functions of the loss-prevention manager's job is the training of other managers and their subordinates—essentially training in security and safety for all employees of the business establishment. Although the major concerns are the control and training of loss-prevention security or safety personnel, the responsibility goes further. The loss-prevention manager must also become involved in the training of other coworkers concerning security- and safety-related areas. This will include employment policies and rules of conduct, internal and external theft, and, most importantly, the safety and protection of life and property. One of the more important elements that a company can show on its behalf when threatened with a criminal or civil action is documented evidence of the appropriate and required training of all employees, particularly specific employees who are responsible in the areas of safety and security.

Preliminary Instruction

As required by OSHA (Occupational Safety and Health Administration) standards, the newly hired worker on the first day of employment will be instructed in the following:

- How to react to a fire or similar emergency notification according to the company's fire safety procedure
- His or her "right to know" (hazard communication) concerning chemicals and other hazardous substances on premises, and the use and location of material safety data sheets (MSDS)

Orientation Training

Along with human resources personnel, the loss-prevention manager should be involved in the initial orientation of new employees. This provides the loss-prevention manager the opportunity to meet all new hires and stress the importance of security and safety in the facility.

Introduction

The loss-prevention manager should introduce himself or herself to the new employees and also have them introduce themselves and provide a little information on their background. Discuss the makeup of the loss-prevention department, what duties are conducted by the department, and company policy regarding violation of company rules and regulations as well as crimes that might be committed by employees. Additionally, discussion should also include the role that all workers play in security and safety.

The Role of Loss Prevention

The basic role of loss prevention for the company should be explained. Discuss the scope of the security operation and the nature of the business. Stress the professionalism of the operation and its members, and the need to have the cooperation of each worker in order to fulfill these responsibilities.

Protection of Life

Explain that there is no greater responsibility than the protection of coworkers, customers, and visitors. This will include not only protection of life, but protection from injury.

Protection of Property

Explain the responsibility of protecting the building, the equipment contained therein to operate the business, proprietary information, and any merchandise or warehouse stock.

Prevention of Loss

Loss prevention is responsible for preventing losses other than losses due to poor products or merchandise or operating decisions. Depending on the business enterprise or the role of the institution, the areas of responsibility include security and safety by routine auditing and inspection, employee and other personal accident investigations and claims, internal and external fraud and larcenies, product liability investigations and claims, building security, burglary, robbery, fire, and fire safety.

Explain how critical it is to the success of the security and safety program that employees advise loss prevention whenever they witness or suspect any suspicious situation or possible violation. All employees should understand the impact of a business disruption due to a criminal act, fire, or any other occurrence that may affect their lives and livelihood.

Accidents

All new hires will be advised on how to handle an accident, what to do, and how to report it. This includes both employee and other personal accidents. They will also be advised that, in the case that an employee is exposed to some human blood or body fluids, that individual falls under the exposure control plan for bloodborne pathogens. Additionally, new hires must be instructed on what not to say or do in the presence of an injured party or witnesses.

Other Training

All new employees should be advised that they will be assigned shortly to other training required by company policy and OSHA standards. This may include:

- Fire safety
- Hazard communication
- Other required safety training contingent on the employee's job description

Subsequent training may also include:

- First-aid/CPR training for first-aid responders or certain job categories
- Bloodborne-pathogen training if the employee falls under the exposure control plan.

Other Issues

Coworker Conduct

Explanation must be given as to the policy and rules of conduct issued at the time of hire. Discuss the importance of reading this information and understanding the rules of conduct in particular. Encourage new hires to ask human resources or loss prevention if they have any questions about the rules and regulations or the policies and procedures.

Violence in the Workplace

All employees should be trained in recognizing aberrant behavior. They should be advised on how to report threats of violence, an assault or any type of harassment and intimidation, and how such reports will be acted upon by the company. The legal and physical ramifications and business issues should also be covered in some detail, along with the duties and responsibilities of the threat assessment team, and the sanctions that an offender will face.

Robbery or Assault by Disorderly or Disgruntled Persons

All employees should be instructed on what to do in the event of a robbery or an assault by a customer, visitor, or unknown person. Be certain that everyone understands never to resist anyone with a weapon or a perceived weapon. Employees must cooperate fully. In the case of a disorderly or disgruntled person who may be causing a scene, advise employees to request a security officer immediately.

Closed-Circuit Television (CCTV)

Explain where cameras are located, under what legal provisions that cameras are operated under, and the purposes for which they are used.

Safety Training

The loss-prevention manager will be responsible for ensuring that the following training classes are conducted:

Pretraining: Preliminary safety instruction for employees upon hire (OSHA required).

Fire safety: Required of all coworkers as per company policy or the OSHA standard, soon after hire.

First-aid/CPR training and certification: Required of all first-aid responders; may include other personnel as needed; outside training and certification will be provided.

Hazard communication training: Required of all coworkers as per the OSHA standard; includes annual retraining.

Bloodborne pathogens training: Required of all first-aid responders, or based on occupational exposure, as per the OSHA standard; to include instruction in the exposure control plan and hepatitis A, B, and C; introduction only to coworkers as part of their hazmat training; includes documentation for hepatitis B vaccination required of those coworkers noted in the exposure control plan.

Back safety instruction: As may be required for those employees whose work exposes them to possible back injury.

Loss-Prevention Training

The loss-prevention manager shall conduct initial training in security and safety for all newly hired loss-prevention officers and ongoing training of all officers on a routine basis. This training will include the following:

- Basic first aid and CPR certification
- Delineation of duties and responsibilities in response to an emergency and/or an disaster affecting persons and property
- Pertinent local and federal laws
- Investigative procedures and techniques
- Techniques in the observation of apparent or possible criminal activity
- Interview and interrogation techniques
- Arrest laws of the state in which employed
- Apprehension, restraint, and detention procedures
- Procedures for dealing with local law enforcement and other investigative agencies, banks, credit card companies, insurance carriers, and their investigators
- The writing of a good report
- Proper use, compilation, and filing of required forms
- Accident investigation and follow-up
- Accident and statistical analysis
- Review of alarm systems and building procedures
- Review and acknowledgment of company's safety and security policies and procedures
- Case preparation and court testimony
- Legal and proper use and installation of CCTV and other surveillance equipment
- A working knowledge of the fire command station or fire board, the burglar alarm system, and the duties attendant to these areas

The Safety Committee

An effective safety committee is the backbone of every successful safety program if the committee (a) adequately promotes active interest in accident prevention and (b) serves as a means of safety communication to and from the work force. Other than the reduction of liability concerns, this committee can be very useful in informing the management of safety and security issues before they become more serious.

One of the best ways to maintain this high level of interest is to involve employees in the safety committee. Membership should be rotated periodically so that all or most employees get a chance to serve. In order for a safety committee to be effective and feel that its function is important, top management should actively participate and serve on the safety committee.

Goals of the Safety Committee

There are several specific goals that a safety committee should identify and promote:

1. ***Encourage support for the safety programs***: Supporting the safety program also means that each individual member of the committee should continually set a good example for others and feel free to evaluate and comment on all facets of the safety program, both within their own departments and the entire company. "Support" in the sense applied here really means a full-time, genuine concern and involvement in the safety of fellow employees and visitors.
2. ***Create and maintain coworker interest in safety***: There are two primary means available to help control those unsafe acts that are largely responsible for most accidents:
   - Education of coworkers
   - Development of safety awareness in each employee. Creating and maintaining coworker interest or safety awareness can be done in a number of ways. Safety committees are responsible for evaluating existing or potential safety awareness programs and implementing safety programs and training as may be required.
3. ***Further the education/training of coworkers in safety***: The initial training and retraining of coworkers is a management responsibility. However, safety committees can serve as the "ears" of management in determining when additional or refresher training might be needed to help avoid accidents. Permanent members should receive specialized training on general topics of safety.
4. ***Inspect for and correct unsafe acts or hazards***: Prompt identification and correction of unsafe conditions is a must. However, identification of unsafe conditions is only the beginning. Unsafe conditions are estimated to be the primary causes in only 10–15% of such cases. Unsafe acts are usually the primary causes and the sole causes in 85–90% of all accidents. If a safety program is to be effective, it must focus on unsafe acts. Therefore, safety committees must continually be alert and periodically review possibly unsafe behavior of coworkers.

- There are several more issues with which the safety committee should be concerned. Safe layouts; safe displays; emergency planning; proper maintenance and handling of tools, equipment, and machinery; and the protection of employees are a few that the committee should consider as within its area of oversight. Keep in mind, however, that management retains accountability over all of these areas.

5. ***Foster communication between coworkers and management***: Rarely is anything accomplished without the cooperation of someone else. This certainly is true with regard to safety. Managers and coworkers must cooperate with each other. The safety committee plays a large part in effecting the spirit of cooperation by assuming the responsibilities enumerated above.
6. ***Communicate***: Although the need for communication seems basic and evident, it has been a stumbling block to the effective functioning of many safety committees. When an existing deficiency is detected in the safety program, it must be directly communicated to the right people.
   - It is all too easy, however, to become so concerned over the mechanics of operating a safety committee that we lose sight of where we are going. The committee, and especially the chair of the committee, should periodically step back from the day-to-day considerations to review the committee's activities in light of its established goals and responsibilities. Doing so can result in a more effective and efficient committee.

Function of the Safety Committee

The function of the safety committee is to create and maintain a high level of interest in the awareness of safety among employees at all levels. To do this, the committee should perform the following:

- Provide a means of communications concerning safety matters between management and the work force
- Develop safety policies and recommend their adoption by top management
- Create and maintain an active interest in accident prevention
- Identify unsafe work practices and conditions and suggest appropriate remedies
- Encourage feedback from all levels in all areas of the facility regarding problems, ideas, and solutions relating to safety and health
- Assist in accident investigations and develop recommendations to eliminate the accident cause
- Recommend and assist in the development of specific safety programs
- Assess safety equipment needs
- Disseminate safety policy material
- Identify specific safety-related problems that seem to be recurring and develop appropriate preventive measures

Establishment of a Safety Committee

The safety committee should be representative of the various departments throughout the company or facility. The members should be knowledgeable and in tune with operations within their own departments. Department managers and supervisors should be included as members. Membership on the committee should be rotated at least on a semiannual basis, with the rotation structured so that several experienced members remain on the committee at all times. The control of the safety committee is a management responsibility. Therefore, the operations manager, human resource manager, or the loss-prevention manager should serve as the chair of the committee.

The following describes a representative safety committee:

- Permanent members
  - Operations manager, facility manager, or senior administrator (chair), or institution, hospital, or school administrator (chair)
  - Loss-prevention/security manager/risk manager (as applicable)
  - Fire safety director or senior fireguard (as applicable)
  - Human resources manager
  - Maintenance manager or manager of the engineering department (as applicable). If these managers are unable to attend a particular meeting, a deputy or assistant should be assigned.
- Other members
  - A worker, manager, or supervisor should represent the following departments on a rotation basis:
    - Goods flow, stock control, warehouse
    - Office personnel: financial, records, etc.
    - Maintenance/engineering
    - Housekeeping/janitorial, food services
    - Decorators, painters, carpenters
    - Customer service showrooms
    - Medical, school, institutional, or administrative departments

Selection of Committee Members

All prospective members of the safety committee should be aware of the function of the committee, and selection should be based on the following attributes:

- ***Willingness*** to serve and carry out the committee functions.
- ***Leadership***: Chosen members should have the respect of their peers, subordinates, and superiors, and they should have demonstrated the ability to gain the respect of new people they work with.

- ***Technical skill***: It is best to choose members who have demonstrated safety-related skills; this may include first-aid responders, certified EMTs, volunteer firefighters, building construction workers, members of the police auxiliary, etc.

Safety Committee Meetings

Safety committees are only as effective as (a) their meetings and findings and (b) management's commitment to the security and safety process. Conducting a successful safety committee meeting should be the highest priority. If meetings are too formal, they can stifle creative ideas, meaningful decisions, and most importantly, member interest and participation. On the other hand, if they are too informal, they will achieve little.

To keep the meeting on track and meaningful, a basic agenda must be developed to address topics appropriate for each meeting. Good meetings require discipline and allow only information related to the topics on the agenda to be discussed. If a meeting is open to general discussion, the meeting could last indefinitely, with little accomplished. Set a definite time limit for meetings. Forty-five minutes is common for an effective meeting, with a maximum of one hour. All members should be reminded of the time limit and to keep their comments to the topic under discussion.

Safety committee meetings should be held monthly and there should be no cancellations. If meetings are canceled, the importance of the committee and the meeting will be diminished.

These meetings should cover employee and client/customer/visitor accident review, inspections and audits, fire and safety violations, near misses, and safety education.

The following is an example of the order of business to be conducted:

1. Outstanding issues discussed at the last committee meeting
2. Internal inspections and audits, violations, and problems
3. Fire code violations
4. Overview of employee and client/customer/visitor accidents
5. Mechanical problems or violations as applicable: forklifts, trucking, entrances and exits, access control, controlled or secured spaces, parking, unsafe conditions
6. Specific safety, security, and health concerns brought forth by each committee member

Minutes of the meeting should be taken and maintained. Copies of the minutes should be distributed to all members of the committee and to all department heads, and another copy of the minutes should be posted on a bulletin board so that all employees may review the business of the committee.

All company employees should be encouraged to approach a safety committee member assigned from their department with any ideas or problems they may view as of interest to the committee.

Occupational Exposure to Bloodborne Pathogens (29 CFR 1910.1030)

Exposure Control Plan

The following information may be used as an effective exposure control plan that will satisfy the mandated requirement, and may be modified to fit a particular business establishment.

> Identification of job classifications or tasks where there is a possible or expected exposure to blood, body fluids and other potentially infectious substances.

Those employees who have occupational exposure to bloodborne pathogens include those who provide first-response medical care and face a reasonable expectation of contact with blood or other infectious materials.

- *Group One*: first-aid responders, health care workers: doctors, dentists, nurses, technicians, attendants, etc.
- *Group Two*: restaurant employees, whose daily tasks include the use of knives and sharp tools and who handle food; matrons, child- and adult-care employees, and housekeeping staff who may come into contact with blood, blood products, body fluids, urine, or feces.

Note that those employees not designated or described in Group One and Group Two—but give aid to a visitor, client, student, customer, or coworker as a "good Samaritan"—are not covered by the training and vaccination requirements of this law. However, they are covered by the *evaluation of exposure incident regulation* contained herein.

> Any security or safety officer or *any other employee*, who has been exposed to blood and/or body fluids that may have entered his or her system, is covered under this exposure act.

The employer is required to follow the program as outlined in the section entitled "Procedure to Follow if an Exposure Incident Occurs."

Communicating Hazards To Employees

Communication and/or training will be given at no cost to employees who fall under Group One or Group Two noted previously.

- Such communication and/or training will be presented at time of initial assignment.
- Further communication and/or training will be given following the initial training at least once a year and be documented. Also, the employee will be trained when existing tasks are modified or when assigned new tasks that involve occupational exposure to bloodborne pathogens.
- Persons conducting such communication and/or training must be knowledgeable about the subject matter, particularly as it relates to emergency response personnel.
- Information provided must be appropriate to the literacy, language, and educational level of the audience.
- A question-and-answer period shall follow all training.

Training Program

A training program shall consist, at a minimum, of the following elements:

- An accessible copy and explanation of the regulatory text
- A general explanation of the epidemiology and symptoms of bloodborne diseases
- An explanation of the modes of transmission of bloodborne pathogens
- An explanation of the written exposure control plan and how to obtain a copy if not given to the employee at time of training
- An explanation of how to recognize events that may involve exposure to blood and other potentially infectious materials
- An explanation of the basis for selecting personal protective equipment, including information on the types, selection, proper use, location, removal, handling, decontamination, and disposal of personal protective equipment
- An explanation of the use and limitations of safe work practices, engineering controls, and personal protective equipment
- Information on hepatitis B vaccination such as safety, benefits, efficacy, and availability to the employee
- An explanation of the procedure to follow if pathogen exposure occurs, including methods of reporting and the medical follow-up that will be made available
- Information on the postexposure evaluation required in the event of an exposure, and information on emergencies that relate to blood or other potentially infectious materials, follow-up procedures, and medical counseling

- An explanation of the information found on warning signs, labels, and color-coding

Appropriate bloodborne pathogen training should be administered either by the company or the housekeeping contractor, particularly if employees fall into Group Two as described previously. However, all other compliance with this standard pertaining to contractual housekeeping employees shall be the responsibility of the contractor and/or employer.

Preventive Measures

Hepatitis B Vaccination

The company shall make available, free of charge, at a reasonable time and place, the hepatitis B vaccine and vaccination series to all employees who are at risk of occupational exposure (Group One and Group Two, as defined previously). Booster doses recommended by the U.S. Public Health Service shall also be provided. The company shall offer this vaccine and vaccination series after initial training and within 10 working days of initial assignment unless

- The worker has previously received the complete hepatitis B vaccination series
- Antibody testing reveals that the worker is immune
- Medical reasons prevent the employee from being vaccinated

All medical evaluations and procedures must be performed by or under the supervision of a licensed physician or an appropriately trained and licensed health care provider, and administered according to current recommendations of the U.S. Public Health Service.

The company cannot require an employee to participate in an antibody prescreening program to receive the hepatitis B vaccination series.

The company shall provide vaccinations at any time, even though the employee may initially decline such vaccination but later requests and/or accepts treatment.

Employees who decline the vaccination must sign a declination form that will be provided at the training session. If the employee wishes, other arrangements for vaccination will be made by the human resources office.

Documentation of vaccination or signed declination forms will be maintained as required and described below under "Record Keeping".

Universal Precautions

Universal precautions constitute methods of infection control in which all human blood and certain human body fluids are to be treated as if known to be infectious for HIV, HBV, and other bloodborne pathogens.

Universal precautions are to be observed in all situations that present a potential for contact with blood or other potentially infectious material. Under circumstances in which differentiation of body type fluids is difficult or impossible, all body fluids are to be considered potentially dangerous.

Methods of Control

To the extent feasible, the company will institute the following controls to eliminate or minimize employee exposure to bloodborne diseases:

Engineering Controls

- Reduce employee exposure in the workplace by either removing the hazard or isolating the employee from exposure. This would include proper containers for the disposal of infectious or contaminated materials.
- The company will provide disposable devices for first-aid responders, as noted under "Personal Protective Equipment."

Work Practice Controls

Alter the manner in which the task is performed.

- The company will ***correctly*** dispose of used bandages, gauze, and other emergency items that may contact blood or other potentially infectious materials.
- Correct sanitization procedures will take place for any area that may have been contaminated as soon as practical.
- The company will provide readily accessible hand-washing facilities and ensure that personnel wash hands and any other exposed skin area with soap and water, and flush mucous membranes with water immediately following contact with blood or other potentially infectious material, or after removing personal protective equipment.

Personal Protective Equipment

OSHA standards require appropriate personal protective equipment so as to reduce work risk of exposure.

- The company will provide eye and face protection, surgical rubber or latex disposable gloves, and disposable CPR devices for use by first-aid responders.

- The company *may* also provide to loss-prevention officers cut/stick-resistant gloves that shall be used to search detained subjects.

Housekeeping Procedures

- The company will ensure that all contaminated equipment and work surfaces shall be decontaminated with a disinfectant upon the completion of the emergency procedure or when such surfaces become contaminated by splashes, spills, or contact with blood or other potentially infectious materials.
- Broken glassware, which may be contaminated, must be picked up *only* by mechanical means such as tongs, brush and dust pan, or forceps, and never with bare or gloved hands.
- Color-coded, leak-proof, and impervious bags or containers will be provided for contaminated waste.

Minimizing Exposure

The company will attempt to ensure that employees observe the following precautions for handling and using personal protective equipment:

- Remove garments penetrated by blood or other infectious materials immediately, or as soon as feasible.
- Before leaving the work area, contaminated protective equipment shall be placed in designated areas or containers for storing, washing, decontamination, or discarding.
- Wear appropriate gloves when there is a potential for hand contact with blood, other potentially infectious materials, mucous membranes, and non-intact skin, and when handling or touching contaminated items or surfaces.
- Disposable gloves shall be replaced as soon as practical when contaminated, torn, punctured, or when their ability to function as a barrier is compromised. They shall never be reused.
- Appropriate face and eye protection shall be worn to protect the eyes, nose, or mouth from splashes, spray, splatter, or droplets of blood or other potentially infectious matter.

Procedure to Follow if an Exposure Incident Occurs

- An ***exposure incident*** is any contact by the eye, mouth, other mucous membrane, non-intact skin, or parenteral contact with blood or other potentially infectious material arising from employment.
- All employees shall ***immediately report exposure incidents***. This allows for timely medical evaluation and follow-up by a health care professional, as well as for timely testing of the source individual's blood for HIV and HBV.

- The company will treat all reports and incidents involving an employee as strictly confidential.
- A company employee accident or similar report will be compiled and submitted on ***all reported exposure incidents***. Thorough assessment, confidentiality, and timely evaluation will be considered critical.
- The company will direct the employee to a health care professional as soon as practical after exposure.
- The company will provide the health care professional with a copy of the Bloodborne Pathogen Standard, a copy of the coworker's job description as it relates to the incident, a copy of the accident report including the route of exposure, and relevant employee medical records, if available, including hepatitis B vaccination status.

Medical Evaluation

The medical evaluation and follow-up must, at the very least:

- Document the details and routes of exposure.
- Identify and document the source individual if feasible and if not prohibited by law.
- Obtain consent and test the source individual as soon as possible to determine infectivity, and document blood test results. (Testing cannot be done in most states without written consent. If consent is not obtained, the company must show that legally required consent could not be obtained. Where consent is not required by law, the source individual's blood should be tested and the results documented.)
- If the source is known to be infectious for HIV or HBV, testing need not be repeated to determine the known infectivity.
- The company shall provide the exposed employee with the test results and information about applicable disclosure laws and regulations concerning the source identity and infection status.
- The company shall obtain consent, collect, and test the blood of the exposed employee as soon as possible after the exposure incident.
- If the exposed employee consents to baseline blood collection but does not consent to HIV serologic testing, blood samples must be preserved for at least 90 days. If, within 90 days of the exposure incident, the worker agrees to have the baseline sample tested, such testing shall be conducted as soon as feasible.

Following the postexposure evaluation, the health care professional will provide a written opinion to the company. This opinion is limited to a statement that the worker has been informed of the results of the evaluation and told of the need, if any, for further evaluation or treatment. All other findings are confidential. The company must provide a copy of the written opinion to the worker within 15 days of the evaluation.

Record Keeping

Employee medical records and training records shall be maintained.

Medical Records

A confidential medical record for each employee with potential for exposure must be maintained by the company according to OSHA's rule governing access to employee exposure and medical records (Title 29 CFR, Section 1910.20[e]).

Additionally, medical records shall include the following:

- The employee's name and social security number
- The employee's hepatitis B vaccination status, including dates of all hepatitis B vaccinations and any medical records relating to the employee's ability to receive vaccinations (This will include any declination form signed by the employee, if any.)
- Results of examinations, medical testing, and postexposure evaluation and follow-up procedures
- The company's copies of the health care professional's written opinion on examination, testing, and evaluation of the exposed employee
- A copy of any information provided to the health care professional by the company

> Medical records of an employee who has been exposed or has had a potential exposure to blood or a bloodborne pathogen must be kept confidential and maintained for at least ***the duration of employment plus 30 years.*** (See "Transfer and Continuity of Records" below.)

Training Records

The company shall be required to maintain and keep accurate records for 3 years, which will include the following:

- Training dates
- Content and summary of the training
- Names and qualifications of the trainers
- Names and job titles of the trainees

Upon request, both medical records and training records must be made available to NIOSH and OSHA. Training records must be available to coworkers or coworker representatives upon request. Medical records can be obtained only by the coworker or anyone having the coworker's written consent.

Transfer and Continuity of Records

If the company, as the employer, ceases to do business, ***all medical and training records*** concerning employees exposed as described in this standard must be transferred to the successor employer for the mandated 30-year period. If there is no successor employer, the company must notify NIOSH for specific directions at least 3 months prior to intended disposal.

> *This sample Exposure Control Plan summarizes the requirements of the OSHA standard (Title 29 Code of Federal Regulations [CFR, Part 1910.1030, eff. 3/6/92]), in that this standard shall apply to every employer with one or more employees who can reasonably be expected to come into contact with blood and other specified body fluids in carrying out or performing their duties. It further requires an employer to prepare a written exposure control plan and mandates that the plan evaluate routine tasks and procedures that involve exposure to blood or other potentially infectious materials, identify workers performing such tasks, and furnish a variety of methods to reduce risks in an effort to provide safe working conditions and an environment that protects workers from exposure to health hazards.*

Hepatitis A, B, and C

Hepatitis is characterized by inflammation of the liver. Viral hepatitis refers to several common diseases caused by viruses that can lead to swelling and tenderness of the liver.

The most common types of viral hepatitis are hepatitis A, hepatitis B, and hepatitis C. There are other forms of hepatitis that are less common; these include hepatitis D and E, as well three other lesser known viruses. Vaccines exist for Types A, B, and D, but there is no vaccine for Type C, the most serious of the types.

Hepatitis B and C can lead to serious, permanent liver damage, and both are causes for concern, particularly to those employees who may come into contact with bloodborne pathogens.*

* The information in this section is derived from the American Liver Foundation, 1996.

Hepatitis A (Infectious Hepatitis)

Hepatitis A virus (HAV) is highly contagious and is contracted by eating food or drinking water contaminated with human excrement. The Centers for Disease Control and Prevention (CDC) estimates that 150,000 people in the United States are infected each year by hepatitis A. Acute hepatitis A causes sudden illness and symptoms similar to the flu, including jaundice. This is a mild type and usually resolves itself within six months, is almost never fatal, nor does it develop into a chronic disease. The virus is present in fecal matter, and outbreaks easily occur in child day-care centers, infected food contaminated by food handlers, and shellfish taken from contaminated waters. Sexual contact, day-care employment, infected food handlers, recent international travel, and contaminated needles are the major known risk factors as listed by the CDC.

Hepatitis E resembles Hepatitis A, but is usually a mild form and does not lead to a chronic infection, although a prolonged case of jaundice could occur.

Hepatitis B (Serum Hepatitis)

Hepatitis B virus (HBV) can cause a serious form of infection. This disease is much more prevalent than HIV, the virus that causes AIDS. It is also transmitted in the same manner as HIV (sex, needles, blood or fluid contact), since the virus is present in blood and other body fluids, but it is more contagious. It causes flu-like symptoms but is slower to develop and becomes much more severe. Sometimes, there are no symptoms of illness. An estimated 1.2 million Americans are currently chronic carriers of HBV. Hepatitis B may develop into a chronic disease in up to 10% of the 200,000 newly infected people each year. If left untreated the risk of developing cirrhosis (scarring of the liver) and liver cancer is increased in patients with chronic Hepatitis B. Health care personnel and first-aid responders should be immunized. Vaccination for hepatitis B is highly recommended if there is the possibility of an exchange of blood or body fluids from another person.

Hepatitis D can only occur in those infected with hepatitis B. It is a more severe form of the virus and can be acute or chronic. Many victims develop cirrhosis, and most die of liver disease.

Hepatitis C (non-A, non-B Hepatitis)

Hepatitis C virus (HCV), once known as non-A, non-B hepatitis and identified only within the last 20 years, develops into a chronic infection in up to 85% of the 150,000 newly infected people each year. This type is the most serious of the hepatitis viruses and can take decades after the initial infection to develop. Currently, in the United States, approximately 3.5 million Americans are chronically infected with HCV. Like chronic hepatitis B, if left untreated, the chronic form of HCV has a greater chance of resulting in cirrhosis, liver cancer, or even liver failure. Liver failure due to chronic hepatitis C infection is the leading cause of liver transplants in the United States. Transmission is by blood-to-blood contact, and it is most prevalent among drug users.

The Risk of Infection

People at risk of Hepatitis B or C include health care workers, first-aid responders, people with multiple sex partners, intravenous drug users, and hemophiliacs. Anyone who has had a tattoo, body piercing, or a blood transfusion (prior to routine screening of donated blood that began in 1972 for hepatitis B and in 1990 for hepatitis C), and those who are in close household contact with an infected person, are also at higher risk. Hepatitis B or C can even be transmitted by sharing toothbrushes or nail files contaminated with infected blood. Although these forms of transmission rarely occur, the various ways of transmission for hepatitis C are not fully understood. Moreover, approximately one-third or more of hepatitis A, B, and C cases result in *unknown* sources. This means that you do not necessarily have to be among high-risk groups to become infected with these viruses.

Most people who get hepatitis B or C have *no recognizable signs or symptoms.* They appear perfectly healthy yet still be infected with the disease—and infect others. But some people do experience flu-like symptoms, such as loss of appetite, nausea and vomiting, fever, weakness, tiredness, as well as mild abdominal pain. Less common symptoms are dark urine and yellowing of the skin and eyes (jaundice). The only way these diseases can be positively identified is through blood tests.

If you suspect that you have hepatitis or have been in contact with an infected person or object, consult your physician as soon as possible.

Testing

There are specific tests to identify viral hepatitis A, B, and C. Blood banks also screen donated blood for hepatitis B and C, and will notify you if you have tested positive. The hepatitis A test, if positive, indicates a recent infection, or that you have developed an immunity to the virus due to a prior infection. The tests for hepatitis B can identify:

- Whether you are infected
- Whether you are recovering from the disease
- Whether you have a chronic infection
- Whether you are immune to hepatitis B

The tests for Hepatitis C can show whether you are infected with the virus or were infected in the past. If you do test positive for Hepatitis B or C, there is treatment available that *may* be helpful. If you think that you may be infected, be sure to consult with you physician immediately.

Prevention

Hepatitis A and B can be avoided. One should always practice safe sex and never share objects such as needles, razors, toothbrushes, nail files, and clippers. When getting a manicure, tattoo, or body piercing, make sure sterile instruments are used. Those who are exposed to blood in their work, such as health care workers, first aid responders, laboratory technicians, dentists, surgeons, nurses, emergency service workers, firefighters, police officers, paramedics, military personnel, or those who live with an infected individual, should be vaccinated against hepatitis B. You should also consider being vaccinated for hepatitis A if you work at a day-care center, come in close contact with someone who is infected, travel to geographic areas that have poor sanitation, or live in an area where there has been a recent outbreak of hepatitis A.

Management is required to conduct bloodborne pathogen training for all first-aid responders and all those employees included in the Exposure Control Plan as per the OSHA standard. The standard also requires that all employees included in the Exposure Control Plan should be vaccinated for protection against hepatitis B (HBV). If an employee declines vaccination and/or testing, a declination form must be signed and maintained in the personnel file.

Vaccination and/or testing are to be provided by the company free of charge to all employees covered by the standard. If an employee at first declines the vaccination and/or testing, the employee may request such medical treatment at any time.

Hazard Communication Standard (29 CFR 1910.1200)

Hazcom: The "Right to Know" Law

The following is a brief description of the hazardous material standard and may be used as part of the mandated training program.

"Right To Know" Training

The employee has the right to know the names and hazards of all chemicals or harmful substances that he or she may be exposed to in the workplace.

A workplace notice listing the employee's rights under the standard Right to Know Act shall be posted with other workplace notices at a location viewable by all employees.

The company safety program will follow the OSHA standard based on the following requirements:

Written Plan

A written plan will be made part of the Emergency Procedure Plan. It will include the following:

- A written policy detailing the company's commitment to a safe working environment
- A description of the training required for employees regarding hazardous substances on the premises
- The documentation practices for inventory, training, and procedures
- A description of the labeling system
- A description of the procedures required to monitor and upgrade the written plan
- A method of warning outside contractors of the hazardous chemicals and substances at the site

Inventory

All chemicals and hazardous substances on premises will be identified and inventoried. This list will be categorized by work area. The inventory will be reviewed and updated yearly or as needed. Any new items not on the list will be added and their material safety data sheets (MSDS) obtained.

MSDS

Each facility shall maintain a material safety data sheet (MSDS) on file for all chemical and hazardous substances listed on the inventory. MSDSs will be provided by the manufacturer, importer, or supplier. If not received at delivery, the MSDS will be requested from the proper source.

Labeling

The company will ensure that all containers holding hazardous substances and chemicals are properly labeled with the common name, chemical name, and hazard warning. The labels will follow the OSHA standard. If the container has no label or an improper label, the company should attach the correct label to the container. If substances are transferred to smaller containers, the smaller containers must also be labeled.

Documentation Practices

It is important that inventory, procedures, and training programs be documented. Training documentation may be contained in an employee's personnel file. Documentation is required and must be accessible for inspection by OSHA.

The Right to Know Act covers the employee's rights when working with hazardous substances. Many companies use products that are classified as hazardous. These include everyday items such as paints, cleaners, solvents, cleaning fluids, etc. Approximately one in four workers is exposed to hazardous chemicals in the workplace. It is estimated that 500,000 chemical products exist today, and hundreds more are introduced each year.

All companies are required to conduct an inventory of hazardous substances found in the workplace. A hazardous substance survey form must be compiled and posted in an area where all notices are normally posted.

Hazcom Training

All business facilities must follow the OSHA standard for an annual education and training program for all workers who may be exposed to hazardous substances in their workplace.

The company must be committed to providing a safe working environment for all employees, and provide a written training program for all employees who may be exposed to hazardous substances. The employee is required to participate in the training program at least once a year as required by law, and review all information that he or she should know concerning the hazardous substances in the work area.

The Occupational Safety and Health Administration (OSHA), a federal agency, has set a standard to ensure that all chemicals produced by or brought into a location are evaluated and that such information concerning their hazard is transmitted to all employees. OSHA further ensures that all employees have the "right to know" what chemical, toxic, or other harmful substance are present and the hazards they may cause. The company must provide annual training to all employees, no matter how long employed. Additionally, this training must be fully documented and noted in the personnel file for future reference.

Material safety data sheets (MSDS) will be explained as to their purpose and use. MSDSs contain such information such as:

- Manufacturer's name, address, emergency telephone numbers
- Product name
- Chemical and common name of product
- Characteristics
- Hazards
- Health effects
- Handling and protective equipment required
- First-aid instructions
- Waste disposal methods

Maintenance and Availability of MSDSs

The company will maintain a file of material safety data sheets (MSDS) that shall be accessible to all employees. OSHA suggests that the sheets be placed in an ***identifiable bright yellow plastic loose-leaf binder***. This binder should be located next to a time clock at the staff entrance or other location easily accessible to all employees. An employee may view these sheets at any time, and may ask a loss-prevention or safety officer for more information. The sheets should be categorized and maintained by department. For example:

Administration
Decorators/renovators
Warehousing/shipping/receiving
Special projects
Maintenance and engineering
Food services/restaurant
Manufacturing/assembly
Medical services
Housekeeping/janitorial services
Sales

For reference purposes, there should also be a file for chemicals no longer used. This file may be maintained in the loss-prevention office, and may also be accessed by anyone who may wish to view this file.

Hazcom training will also include an introduction to the Exposure Control Plan and Procedures required by OSHA for those employees who are at risk of occupational exposure to bloodborne pathogens. Training should cover the procedures to follow if an employee comes in contact with blood or body fluids under any circumstance, and the company's role and requirements as directed by OSHA's Bloodborne Pathogen Standard.

A list of all hazardous substances located on premises and included in the MSDS book should be posted in the vicinity of the MSDS book, and updated as changes occur.

These substances can affect your health if not used properly. The acute and chronic effects of exposure or overexposure to these substances are listed in the material safety data sheets along with the symptoms arising from such exposure and the potential for flammability, volatility, and reactivity.

Employee's Responsibility

The employee has a responsibility to:

1. Attend hazcom training when required to do so
2. Understand how to read a label and MSDS
3. Know what precautions to take when handling hazardous substances, and what to do when the unexpected occurs

Moreover, the employee must be aware of the appropriate emergency procedures in the event of exposure and/or contamination, a spill, leak, fire, or other accident in the work area when it concerns a hazardous substance. This information can be found in the MSDSs. It is the responsibility of all employees to use appropriate protective equipment when handling any hazardous substance, and ensure that the proper conditions exist for their safe use. Protective equipment such as eye protection, gloves, aprons, and boots will be provided by the company at no cost to the employee. When the possibility exists regarding exposure to dust and/or vapor mist, or liquids that may splash into the face, mouth, or nose, protection such as respirators or masks will be provided. If an employee believes that more or proper protection is required when handling a hazardous substance, the employee should contact his or her manager or a member of the safety committee. A list of all members of this committee should be posted for the employees' reference.

Terminology

The type of exposure or incident may be described as follows:

Toxic:	means it is poisonous
Tumorigenic:	causes development of tumors
Mutagenic:	causes mutation
Reproductive toxic:	causes reproductive problems
Flammable:	will catch fire
Explosive:	can explode
Corrosive:	causes visible destruction to human skin
Carcinogenic:	causes cancer

Exposure may occur through ***ingestion, inhalation, or topically*** (on the skin).

Acute effects are adverse effects evident immediately or shortly after exposure. These include, but are not limited to, dizziness, eye irritation, headache, nausea, vomiting, and unconsciousness.

Chronic effects are adverse effects that develop slowly over a long period of time or upon repeated prolonged exposure. These include, but are not limited to, cancer, lung and heart damage, skin rash, boils, etc. (e.g., lung cancer because of long exposure to cigarette smoking or asbestosis).

A hazard may be defined as follows:

A ***health hazard*** is any chemical substance that, when exposed to a person through the skin, inhalation, or ingestion, may produce acute or chronic effects.

An ***environmental hazard*** is any substance, emission, or discharge determined to be hazardous and that poses a danger if released into the environment.

A ***special hazardous substance*** is a substance that poses special hazards to health and safety.

A ***hazardous substance or mixture*** is any chemical that is a physical or health hazard.

A ***physical hazard*** is any explosive or flammable material including combustible liquids and compressed gas.

Health hazards are rated from 0 (minimal) to 4 (severe), and hazardous incidents can or may occur if precautions are not taken.

If an employee believes that he or she has been exposed to a toxic or health hazard, the employee must contact any loss-prevention officer for any medical treatment or evaluation, and documentation of the occurrence.

In any medical emergency, employees must immediately contact, or have someone else contact, the operator at the switchboard or the dedicated in-house emergency number. A loss-prevention officer/first-aid responder will come to the employee's immediate assistance. If the emergency is due to an exposure of a chemical or toxic substance, a quick reference to MSDSs will determine what first-aid treatment must be administered.

Remember that an informed employee is aware of all the possible harmful substances used or contained in his or her work area. Follow directions and be mindful of how these substances must be used or handled. If anyone has any questions or concerns about a substance, contact loss prevention.

Response to an OSHA Inspection by the Loss-Prevention Department

The following procedure applies to OSHA inspection or regulations. There are five reasons that an OSHA team or a compliance officer will request an inspection:

1. Imminent danger to employees
2. A fatal accident or a catastrophic incident where three or more people are injured
3. Reaction or response to an employee complaint
4. High hazard rate; reported injuries (OSHA 200 log) higher than those reported by other companies within the same industry
5. Follow-up inspection from a previous visit

The company will not receive any advance notice of an OSHA compliance officer's inspection; it will be unannounced. The following procedure will be followed when an OSHA team or a compliance officer arrives:

- OSHA *must* announce its arrival.
- The switchboard operator or receptionist will notify the operations manager, the loss-prevention manager, or facility manager immediately.
- If the operations manager and loss-prevention manager are not on premises, they will becontacted and apprised of the visit.
- OSHA personnel may be kept waiting for up to an hour legally while notifications are made to company management. OSHA personnel may be escorted to a conference room or office to await the arrival of appropriate management. An attempt shall be made to shield all workers from any questions by OSHA personnel while they wait for company management.
- Upon arrival of company management, an opening conference will take place where OSHA will advise the reason for the visit, and show a copy of the complaint, if any. Under certain circumstances, and with proper legal advisement, the company may require a court-authorized search warrant for OSHA to enter into the building for the purpose of an inspection and/or investigation.
- The company will specify only what OSHA is to inspect and/or view as per the complaint. Management will accompany the compliance officer or his or her team, and may take whatever route to the inspection site as management may consider appropriate.
- All records are subject to inspection if they are part of the investigation and/or the complaint, and must be produced on demand to the compliance officer.
- A closing conference will take place in which OSHA will advise what was found, what requires compliance, and review the number and nature of any violations found. If violations or deficiencies are found, company management will inform the compliance officer of their willingness to correct any problem or act toward an agreeable solution. It is advisable to show good faith to the compliance officer, as it may reduce any penalty assessed. Good faith can be shown in describing the types of routine inspections, audits, and investigative procedures, description of the safety committee, safety committee minutes, safety standards, and the required training in effect for the protection of employees.

Remember not to volunteer information. Answer only what is required. Produce only those records that OSHA requires or has the authority to view.

Appendix C

Homeland Security

Defined

The Department of Homeland Security may be defined as follows:[1]

> Homeland Security is a concentrated national effort to prevent terrorist attacks within the United States, reduce America's vulnerability to terrorism, and minimize the damage and recover from attacks that do occur. Additionally the Department has resources to detect, disrupt and prohibit the movement of weapons of mass destruction and related materials. Moreover, the protection of the American people, critical infrastructures and key resources are to be considered prime incentives.

Today's threat environment involves several factors.

Terrorism

The intent to cause death, injury, and destruction, along with fear, apprehension, and disorder among the populace with the use of weapons of mass destruction (nuclear, radiological, biological, and chemical elements) by a host of groups and individuals—domestic and international—including:

- Left- and right-wing ideologies
- Single-issue groups (white supremacist, environmental, animal rights)
- Religious groups (al-Qaeda, Hezbollah)
- Nationalistic groups (violent Islamic alliances)

Catastrophic National Disasters

Such disasters include

- Infectious diseases with the possibility of pandemics
- Meteorological and geological events, which include hurricanes, tornadoes, floods, earthquakes, volcano eruptions, drought, and famine

Catastrophic Accidents

Such accidents include

- Industrial hazards and events
- Infrastructure failures
- Utility defects and failures (electric, gas, electronic, telephone, computer, etc.)

Critical Infrastructure and Key Resources of Concern

Homeland Security has identified 17 sectors of critical infrastructure and key resources, each intersecting in some manner with physical, cyber, and human elements:

1. Agriculture and food production
2. Banking and finance
3. Chemical industries
4. Commercial facilities
5. Commercial nuclear reactors, wastes, and materials
6. Dams
7. Defense industries
8. Drinking water and water treatment systems
9. Emergency services
10. Energy
11. Government facilities
12. Information technology (including cyber security; the uninterrupted use of the Internet and the communications systems that comprise our cyber infrastructure)
13. National monuments and icons
14. Postal and shipping
15. Public health and health care
16. Telecommunications
17. Transportation systems

The Homeland Security Advisory System

The Level of Threat

The Department of Homeland Security will advise the public of the *level of threat* regarding a terrorist attack.

1. *Condition Green* (low condition): low risk of a terrorist attack
2. *Condition Blue* (guarded condition): declared when there is a general risk of a terrorist attack
3. *Condition Yellow* (elevated condition): significant risk of terrorist attacks
4. *Condition Orange* (high condition): high risk of a terrorist attack
5. *Condition Red* (severe condition): ultimate and severe risk of a terrorist attack

Notes

1. The narrative contained herein has been abridged from "National Strategy for Homeland Security," Homeland Security Council, October 2007 (prepared for the White House).

Appendix D

Web Sites

Web sites are routinely redesigned or eliminated and domain names change. Home pages that are popular today may not exist tomorrow. However, the Web sites listed below are well known, have been in service for some time, and are continually updated to contain current information.

Centers for Disease Control (CDC) www.cdc.gov

The source for information about biological agents, communicable diseases

Centre for Research on the Epidemiology of Disasters (CRED) www.cred/be

Climate Diagnostics Center, USA www.cdc.noaa.gov

Computer Emergency Readiness Team (CERT) www.cert.org

Department of Defense (DOD) www.defenselink.mil/specials/homeland
Links to federal agencies

Department of Energy (DOE) www.energy.gov

Click on "national security"

Department of Homeland Security (DHS) www.dhs.gov

A leading agency under whose aegis many law enforcement and investigative agencies now fall. A variety of agencies, including law enforcement, can be found through this Web site

Department of Labor (DOL)

Occupational Safety and Health Administration (OSHA)

Enforce and administer the laws and regulations governing training of personnel and workplace safety. www.osha.gov

Department of State (DOS) www.state.gov

Travel warnings, crisis awareness, and international issues www.usdos.gov

Department of Transportation, (DOT) www.dot.gov

Terrorism and public transportation

The Disaster Center www.ntis.gov/products

Environmental Protection Agency (EPA) www.epa.gov

Links to "response to 9/11"

Federal Bureau of Investigation (FBI) www.fbi.gov

The leading agency for law enforcement and investigation, including terrorism/counterterrorism

Federal Emergency Management Agency (FEMA) www.fema.gov

A leading agency supporting state and localities in responding to catastrophes and antiterrorism and counterterrorism activities

Food and Drug Administration (FDA) www.fda.gov

Information on defense against bioterrorism and ensuring food safety

Health and Human Services (HHS) www.hhs.gov

Bioterrorism, homeland security, and natural disasters

Institute for Business and Home Safety www.disastersafety.org

Disaster management and resources, property/safety protection and recovery guides

Intergovernmental Panel on Climate Control www.ipcc.org

International Atomic Energy Agency	www.iaea.org.at
Nuclear topics	
Naval Postgraduate School *Prof. Dorothy Denning on cyberterrorism, etc.*	www.nps.navy.mil/da/faculty/dorothydenning/index.htm
Nuclear Regulatory Commission, (NRC)	www.nrc.gov
Nuclear topics	
Ready.Gov	www.ready.gov
Run by DHS; preparing for a nuclear attack; general information on nuclear, chemical, and biological threats	
Red Cross Disaster Relief Agency (Government Publications)	www.redcross.org
The Sans Institute	www.sans.org
The United States Computer Emergency Readiness Team	www.us-cert.gov
World Health Organization (WHO)	www.who/int/pub/en (click on publications)
An agency of the United Nations; biological and general medical information	
Medical advice and information	www.webmd.com
Private Web site	www.disastercenter.com
Information about disasters in the U.S.	

Endnotes

Introduction

1. SBA, "Disaster Preparedness Considerations" (Washington, DC: Small Business Administration, September 2001), http://www.sba.gov/idc/groups/public/documents/fl_miami/fl_mi_disasterbook.pdf.
2. *Business Week Magazine*, May 7, 2007, 73.
3. Ibid.
4. Richard A. Posner, *Catastrophe—Risk and Response* (New York: Oxford University Press, 2004), 15.

Section One

5. *Security Magazine*, January 2002, 16.
6. Ibid.
7. *Security Magazine*, May 2000, 90. Analysis by the Security Group for *SDM Magazine*, based on the 7th annual survey of Fortune 1000 corporate security professionals by Pinkerton Security Services.
8. Anthony D. Manley, *The Retail Loss Prevention Officer: The Fundamental Elements of Retail Security and Safety* (Upper Saddle River, NJ: Prentice Hall, 2004).
9. Ibid.

Section Two

10. International Association of Chiefs of Police, "Project Response: The Oklahoma City Tragedy," http://www.theiacp.org/documents/pdfs/Publications/OklahomaCityTragedy.pdf. An excellent study on threat analysis, disaster response, protective measures, and critical incident management.

11. *Stability and Support Operations*, U.S. Army Field Manual, Chapter 8: "Combating Terrorism." U.S. Army Command and General Staff College, Fort Leavenworth, Kansas. An excellent source on the study of terrorism.
12. Bureau of Public Affairs, U.S. Department of State, Washington, DC.
13. David Alexander, *Conspiracies and Cover-Ups* (New York: Berkley Books, 2002), 130.
14. "Terrorist Attack on Chemical Plants Could Endanger Millions," *Environmental Media Services*, June 20, 2002, www.ems/chemical-plants/zz.02.06.20html.
15. Scott R. Gane, "Security Compliance at Chemical Facilities," *Security Magazine*, July 2007, 32.
16. Ibid.
17. Bill Zalud, ed., "It's in the Mail," *Security Magazine*, September 2007, 80.
18. Ibid.
19. Ivan Eland, "U.S. Ignores Bio-Threat at Its Peril," *Newsday-Long Island*, October 5, 2001, viewpoints section, http://www.independent.org/newsroom/article.asp?id=1224.
20. Bureau of Public Affairs, U.S. Department of State, Washington, DC.
21. Posner, *Catastrophe—Risk and Response.*
22. Manley, *Retail Loss Prevention Officer.*
23. Title 18 USC Section 1831–1839, Espionage Act of 1996.
24. Society of Competitive Intelligence Professionals, http://www.scip.org.
25. John F. Bumgarner, "Computer Security—Wave Goodbye to Liability," *Security Management Magazine*, January 2001, 49.
26. Pete Van DeGohm, "Safeguarding Information—It's Not What You Know," *Security Management Magazine*, September 2001, 96.
27. Lynn Tan, "Four Deadly Security Sins," ZDNet Asia. http://www.zdnetasia.com/news/security/0,39044215,62020417,00.htm.
28. Secure Computing, San Jose, CA.
29. The Sans Institute claims to be the largest source in the world for information security training, certification, and research, www.sans.org.
30. Title 18 USC Section 2701, Unlawful Access to Stored Communications.
31. Alexander. *Conspiracies and Cover-Ups.*
32. Title 18 USC Section 1030, Fraud and Related Activity in Connection with Computers.

Section Three

33. Posner, *Catastrophe—Risk and Response.*
34. Ibid.
35. Ibid.
36. Ibid.
37. Ibid.
38. Title 29 CFR Section 1910.1030, Occupational Safety and Health Standards, Toxic and Hazardous Substances, Bloodborne Pathogens.
39. Title 29 CFR Section 1910.1200, Occupational Safety and Health Standards, Toxic and Hazardous Substances, Hazard Communication.
40. National Fire Protection Association (NFPA), www.nfpa.org (click on learning).

Section Four

41. Small Business Administration, "Disaster Preparedness Considerations."
42. International Association of Chiefs of Police, *Oklahoma City Tragedy.*
43. Ibid., 6.
44. Ibid.
45. Ibid.
46. Ibid.
47. Ibid.
48. Ibid.
49. Ibid.
50. FBI—ANSIR program, ansir@leo.gov. See glossary; may be used to request appropriate intelligence on a routine basis.
51. Ibid.

Section Five

52. Title 29 CFR Section 1900.5(a), OSHA's general duty clause: provides that "each employer shall furnish to each of his employees a place of employment which are free from recognized hazards that are causing or likely to cause death or serious physical harm to his employees."
53. Ibid.
54. NIOSH, "Violence in the Workplace," fact sheet, 1995.
55. NIOSH, "Violence in the Workplace," fact sheet, 1998.
56. Title 29 CFR Section 1900.5(a).
57. Associated Press, "Rage; 25% Feel Angry," *Newsday-Long Island,* August 11, 1999, Business Section. See also NIOSH fact sheet 1995 and NIOSH fact sheet 1998.
58. "Workplace Violence Awareness and Prevention," a position paper by the Long Island Coalition for Workplace Violence Awareness and Prevention, Mineola, NY, 1996.
59. "Violence and Disciplinary Problems in U.S. Public Schools, 1996-1997," March 1998 (NCES 98-030); National Center for Educational Statistics (NCES), U.S. Department of Education, Office of Educational Research and Improvement, Washington, DC.
60. *Diagnostic and Statistical Manual of Mental Disorders; American Psychiatric Association,* 4th ed. (Washington, DC: American Psychiatric Association, 1994).
61. Ibid.
62. NIOSH publication 96-100.
63. Long Island Coalition for Workplace Violence Awareness and Prevention, "Workplace Violence Awareness."
64. Manley, *Retail Loss Prevention Officer.*
65. Title 29 USC Section 158, The National Labor Relations Act (a/k/a The Wagner Act of 1935). See also The National Labor Relations Board (NLRB). See glossary.
66. Manley, *Retail Loss Prevention Officer.*

Section Six

67. Manley, *Retail Loss Prevention Officer*.
68. Gerald N. Hill and Kathleen Thompson Hill, *The People's Law Dictionary* (New York: MJF Books/Fine Communications, 2002).
69. Ibid.
70. Manley, *Retail Loss Prevention Officer*.
71. As contained in the United States Constitution.
72. Title VII of the Civil Rights Act of 1964.
73. The author wishes to note that the term *gypsy*, when used in the narrative herein, does not refer to a particular ethnic group, but rather it refers to a criminal lifestyle and is not intended to describe or include the law-abiding members of the Romani people or their culture.

Glossary

ABC warfare: An acronym for atomic, biological, and chemical warfare; also a class that includes weapons of mass destruction. May also be known as NBC (N = nuclear). See also WMD.

access control: Terminology that covers the management and mechanisms of entry and the capability of people who wish to enter a building or an enclosure or to gain entry into an information system. To access a structure, a hard key, a magnetic or electronic card key, personal identification number (PIN), eye (iris) recognition, handprint, fingerprint, or voiceprint may be utilized. Control of access to information and data via computers may include *authentication* and *encryption*.

agroterrorism: The contamination of the food supply by terrorist actions.

AIA: American Insurance Association, which includes property, casualty, and surety companies. Also includes the Fire and Theft Index Bureau, a department of AIA, which compiles and shares fire and theft claims information in an effort to prevent fraudulent claims. See also PILR.

Al-Qaeda: An international terrorist organization. At the time of this writing, Osama bin Laden is still considered to be the head of this organization.

ANSIR: Awareness of National Security Issues and Response, a program set up and controlled by the FBI via each of its local field offices to apprise the public and private sector of computer and economic espionage and sabotage, counterintelligence, counterterrorism, cyber and physical infrastructure protection, including all national security issues. Further info: www.fbi.gov. Click on ANSIR or your local FBI office, or contact ansir@leo.gov.

anthrax: *Bacillus anthracis disease.* Considered by terrorism experts and the CDC to be in the first rank of potential bioweapons. Usually transmitted topically (on the skin), but if inhaled, death will be the result.

anticrop agent: A biological or chemical agent used to cause disease or damage to food or industrial crops.

arraignment: Bringing a defendant before the court to answer an accusation or complaint.

arrest: Taking a person into custody for the purpose of answering a criminal charge or civil demand.

assault: Any intentional, unlawful offer, attempt, or threat of corporal injury to another by force, or force unlawfully directed to the person of another, under such circumstances as to create a well-founded fear of imminent peril, coupled with the ability to effectuate the attempt if not prevented. An attempt to offer or beat another, without touching the person, as when one lifts up a fist in a threatening manner at another. The touching of another's person, if done willfully or in anger, constitutes battery. Also includes intentionally, recklessly, or with criminal negligence causing physical injury to another person.

assault and battery: Always includes an assault; therefore, the two terms are combined as *assault and battery.* An unlawful touching of the person of another by the aggressor or by some substance put in motion by the aggressor. See also battery.

atomic weapon: A bomb that will produce an exceptionally large explosion and thermal blast, containing radioactive material that, after detonation, will spread over a large geographical area.

authentication: A method that confirms the user's identity, such as a password or PIN; the use of a smart card or ATM-type card; or a fingerprint, iris, etc. The use of multiple methods in combination would provide the strongest authentication.

bacterial agent: A live pathogenic organism that can cause disease, illness, or death.

battery: Intentional and wrongful physical contact with a person without his or her consent that entails some injury or offensive touching. Unlawful beating or unlawful touching of one person by another.

biological agent: A microorganism that will cause disease or death of people, animals, or plants or the deterioration of material. The weaponization of a bioagent is a process in which a harmful biological agent is manufactured so as to ensure stability and predictability so that it can be safely handled by the perpetrator.

biological contamination: The presence of an infectious substance on a human, an animal, or an environmental surface.

biometric: The measurement of physical characteristics such as fingerprints, DNA, and retinal patterns, which is used for verification of the identity of individuals.

bioterrorism: The use of biological weapons as a form of terrorism.

bribery: An offer or receipt of any gift, loan, fee, reward, or other advantage to or from any person as an inducement to do something dishonest, illegal, or that constitutes a breach of trust in the conduct of business.

CDC: Centers for Disease Control and Prevention, the lead federal agency for protecting public health and safety along with the development and application of disease prevention and control.

CERT: Computer Emergency Response Team: coordination center operated by Carnegie Mellon University; federally funded by Department of Homeland Security and Department of Defense as a research center providing computer forensics and security, incident execution, training, research and development, and public advisories. Excellent information source available to the private investigator.

Computer Emergency Readiness Team: Part of the Department of Homeland Security. Analyzes cyber threats and vulnerabilities and disseminates cyber threat information.

CFR: Code of Federal Regulations (federal law).

civil action: Action brought to enforce, redress, or protect private rights. In general, all types of non-criminal proceedings.

civil law: A body of statutory and common law relating to private rights and remedies available to a citizen.

compensatory damages: Those damages arising from a breach or tortious act, and which can be readily proven to have been sustained, and for which the injured party should be compensated as a matter of right. See also *punitive damages*.

contract law: The law that governs contracts, which can take several forms, e.g., a written or a verbal agreement between two or more people that creates an obligation to do, or not to do, something. The agreement creates a legal relationship of rights and duties. If the agreement is broken, then the law will provide certain remedies.

crime: An *act* or *omission* forbidden by law and punishable, upon conviction, by some penalty, such as a fine, imprisonment, or death; it is an offense against the state. An act or omission forbidden by society—the people—through their representatives. As an example, consider that possession of marijuana is a crime today, but it may not be a crime tomorrow. A proscribed crime is whatever society says it is.

criminal trespass: To knowingly *enter* or *remain* unlawfully in or upon private premises or a publicly restricted area.

critical infrastructure: Defined in the U.S. Patriot Act of 2001 (P.L. 107-56, Section 1016e) and adopted by reference in the Homeland Security Act of 2002: "Systems and assets, whether physical or virtual, so vital to the United States that the incapacity or destruction of such systems and assets would have a debilitating impact on security, national economic security, national public health or safety, or any combination of those matters."

cross-contamination: The process by which the handling of a contaminated or toxic substance (as in sorting or delivery) is transferred to another medium so that contamination can occur anywhere along the path.

damages: An award sought by a plaintiff for some wrong (an injury to his or her person, property, or rights) caused by some unlawful act, default, or negligence by another. It is compensation that the law will award for injury done; usually in a monetary form. The award may be to change a practice, to right a wrong, or change a procedure. Damages may be awarded as *compensatory* or *punitive.*

defamation: Damage to the character, reputation, honor, etc., of a person by slanderous and libelous acts. See also *libel* and *slander.*

defendant: The adverse party to an action. In a criminal case, the person who has committed the crime. In a civil action, the person who has committed the wrong and from whom the plaintiff seeks damages.

dirty bomb: A conventional high-explosive bomb surrounded with radioactive material that, when detonated, will spread radioactive fallout over a larger area. This is not to be confused with an atomic weapon.

duress: Any illegal or legal imprisonment used for an illegal purpose, or threats of bodily or other harm, or other means to coerce another and induce him or her to do an act contrary to his or her free will.

Ebola: A virus originating in Africa; a severe and often fatal disease in humans. Three of the four species of the Ebola virus identified have caused disease in humans. The virus liquefies body organs. Mortality rate is between 70% and 90%. See also Marburg and viral agents.

electromagnetic pulse: Electromagnetic radiation from a nuclear explosion. The resulting electric and magnetic fields may couple with electrical or electronic systems to produce damaging current and voltage surges. This phenomenon may also be caused by nonnuclear means.

encryption: A process where coding and decoding are used to secure communications (computer, telephonic) between the receiver and the sender.

epidemic: A disease that is present for a short or limited time in the human or animal population that is transmittable to humans. Mortality rate is very low.

fallout: The descent to earth of radioactive particulate matter from a nuclear cloud.

false arrest: An unlawful arrest; the taking into custody of a person without probable cause.

false imprisonment: Detention of a subject without justification; a false arrest.

FBI: Federal Bureau of Investigation, the major federal law enforcement agency.

felony: A crime punishable by more than one year in a prison up to the death penalty.

fireball: The luminous sphere of hot gases that forms within a few millionths of a second after detonation of a nuclear weapon and immediately starts to expand.

ground zero: The point at, or directly below or directly above, the center of a nuclear detonation.

hazmat: Hazardous material.

hot spot: An area of unusually high radioactivity within a larger area that has low radioactivity.

hybrid weapon: A biological weapon that combines different agents so that any defense is nearly impossible.

injury: Any damage done to one's rights, reputation, or property.

incident report: A report that is compiled and maintained on all events that require a response from loss-prevention officers. This will include the documentation of all facts and circumstances surrounding the occurrence.

intentional tort: A tort or wrong perpetrated by one who intends to do that which the law has declared wrong, as contrasted with negligence, in which the tortfeasor fails to exercise a sufficient degree of care in doing what is otherwise permissible. Also cited as a *willful tort.*

IRC: Insurance Research Council, a nonprofit division of the American Institute for Chartered Property Casualty Underwriters and the Insurance Institute of America. Provides timely and reliable research to all parties involved in public policy issues and in all lines of property and casualty insurance regarding automobiles, homes, businesses, municipalities, and professionals. Conducts a major study each year.

IRSG: Individual Reference Services Group, a lobbying group consisting of 13 of the largest commercial database companies attempting to fend off a government crackdown by introducing self-regulation into the information-brokering business; provides commercial information sevices to help verify the identity of or to locate individuals including missing witnesses, missing or exploited heirs or children, hidden assets, etc.

judgment: Decision or sentence by a court, justice, or other tribunal on the claims of parties to a litigation.

libel: A malicious publication in printing, writing, signs, or pictures tending to blacken the reputation. See also *defamation* and *slander.*

malice: A willful intent to do mischief; ill will.

malicious prosecution: A prosecution that is begun with malice and without probable cause to believe that the charges can be sustained, and which finally ends in failure; a continuation of an arrest by a person knowing it to be false but who still proceeds with the action.

malware: Software capable of capturing operator input and system data, and then delivering the results to another computer via the Internet, thereby covertly gathering personal and proprietary information.

Marburg: Similar to the Ebola virus. Severe and very often fatal, it liquefies the body's organs. See also *Ebola virus* and *viral agents.*

misdemeanor: A crime punishable by more than 15 days (some states) but less than a year in a local or county jail.

negligence: The failure to exercise a degree of care that a reasonable person would exercise given the same circumstances, and thereby causing injury and/or damage. Negligence could be equated to carelessness (negligent tort).

negligence per se: Without any argument or proof as to the particular surrounding circumstances, conduct—whether by action or omission—may be declared or treated as negligence because it is contrary to the law.

nerve agent: Biological agent that is considered lethal. The agent is rapidly absorbed the through the eyes, respiratory tract, and skin. Clothing heavily contaminated with a liquid nerve agent must be removed immediately rather than remain on the skin. Symptoms are runny nose, chest tightness, dimness of vision, and difficulty breathing, followed by excessive sweating and drooling, nausea, vomiting, involuntary urination and defecation, twitching, staggering, convulsions, confusion, headaches, and coma. Immediate treatment is essential, including atropine injection and artificial respiration; otherwise death will result.

NICB: National Insurance Claims Bureau, formerly the National Automobile Theft Bureau. Includes property and casualty insurance companies that have joined to stop crime and vehicle theft. Maintains a computer database.

NIOSH: National Institute for Occupational Safety and Health, a federal subdivision of OSHA.

NLRB: National Labor Relations Board, a federal agency that enforces the National Labor Relations Act.

NOAA: An acronym for National Oceanic and Atmospheric Administration. A government scientific agency focused on conditions of the oceans, atmosphere, weather, and environment.

NRC: An acronym for Nuclear Regulatory Agency. An independent federal agency created to license and regulate nuclear power plants.

NSA/CSS: An acronym for National Security Agency/Central Security Services. A cryptologic intelligence agency of the U. S. government and part of the Department of Defense used to collect foreign communications and global intelligence.

nuclear blast: An intense shock wave and part of the thermal blast caused by a nuclear explosion. Considered to be approximately 50% of the bomb's total effect.

nuclear radiation: Radioactive material created by a nuclear blast. 14% of its total effect: 4% initial and 10% residual.

OSHA: Occupational Safety and Health Administration, a federal agency that oversees and investigates industrial accidents, deaths, and injuries.

pandemic: A disease that spreads throughout an entire country, continent or the world. Mortality rate is usually very high. See also *epidemic.*

perjury: The act of a person under oath who knowingly and willfully swears falsely in a matter material to the issue or point in question. Perjury is a crime.

PILR: Property Insurance Loss Register, a national data registry allowing insurance underwriters to inquire of subjects under investigation for excessive claims or insurance fraud. Also includes a list maintained by the American Insurance Association (AIA) of all fire losses over $500,000, and maintains a database to access information of duplicate coverage or loss. See also AIA.

plague: A disease caused by a bacterial agent. A serious outbreak in the fourteenth century throughout Europe was known as the *bubonic plague*, the black death and caused a manifestation of body lesions. A variation was also known as *pneumonic plague*, which attacked the respiratory system. Historians believe that up to 60% of those infected perished. Infection can be treated.

plaintiff: One who brings a suit, bill, or complaint against another; one who seeks damages for some wrong. A plaintiff may be a person, a group, or a corporation.

prima facie: A case made out of evidence sufficient to counterbalance the general presumption of innocence. Enough evidence to proceed with the action in court. See also prima facie evidence.

prima facie evidence: Evidence showing the existence of the facts and which, if uncontradicted, is sufficient evidence to maintain the proposition affirmed.

probable cause: A conclusion reached by a reasonable person (ordinary intelligence, experience, and judgment) after examining all of the facts and circumstances concerning the question at issue; once it has been determined that an offense did take place, leading a reasonable and prudent person to believe that the offense was in fact committed by a particular perpetrator, an arrest can take place.

product liability: The legal liability of manufacturers and sellers of a product to compensate buyers and users and browsers and bystanders for damages or injuries suffered because of a defect in the goods purchased or for sale.

punitive damages: Compensation, separate and in excess of compensatory damages, which serves as a form of punishment to the wrongdoer who has exhibited malicious, wrongful, and willful conduct.

radiation sickness: An illness caused by exposure to ionizing radiation. Symptoms include nausea, vomiting, and diarrhea, followed by loss of hair, hemorrhage, inflammation of the mouth and throat, and loss of energy.

resource convergence: The responses of police, fire apparatus and personnel, and other public safety personnel and equipment that may cause confusion and chaos by blocking roadway access in their effort to aid, assist, and control the incident. A predicament of this type will impact the responders' ability to assess the nature of the incident, because they may have only limited resources available until complete access is obtained.

ricin: A toxin. There is no vaccine and no cure.

saxitoxin: A toxin. There is no vaccine and no cure.

scope of employment: The actions of an employee that further the business of the employer. The manner or behavior involved in these actions must be within the authority given to the subject by the employer. This concept falls under the statutes and precedents relating to the relationship of an employer and employee. "Let the master answer" (*respondeat superior*) is a key doctrine establishing that an employer is responsible for the actions of an employee (or agent) in the course of employment. If the subject behaves unlawfully, inappropriately, or outside policies and procedures, he or she acts outside of the scope of employment, and the employer can be held free of any liability.

slander: Defamation by words spoken; malicious and defamatory words tending to damage the reputation of another. See also *defamation* and *libel.*

smallpox: A highly contagious disease. The World Health Organization believes that this disease has been eradicated worldwide, but it is also believed that some radical countries have produced and stockpiled the smallpox strain. However, the CDC believes that current vaccine levels are adequate if the populace becomes exposed.

SNS: Strategic National Stockpile; a number of stockpiles have been developed by the U.S. Centers for Disease Control and Prevention throughout the country containing antibiotics, antidotes, and medical supplies and equipment, along with a certain number of controlled substances, for use during a chemical, biological, or radiological exposure or attack.

subpoena: A writ or order requiring the attendance of a person at a particular time and place to testify as a witness.

subpoena duces tecum: Latin for *produce the record.* A writ or order by the court requiring a person to produce a particular document in court or before a legal tribunal.

thermal radiation: The heat and light emitted from a fireball caused by a nuclear blast. It accounts for 35% of its effect.

three lethal rings: When viewing a cross-section of a nuclear detonation, three rings will be observed, all lethal: (1) thermal radiation, (2) nuclear radiation, and (3) the blast.

time record: A time record (or time log) compiled and maintained by a responsible individual present at the scene with access to the comings and goings of all personnel, public safety, and emergency equipment, communications and announcements, individuals and important facts of concern. This record will include times of all notifications and to whom, arrivals and departures of police, fire, medical assistants and ambulances services, all public and safety authorities, all public utilities, company management and anything of importance to the occurrence. This important record is required in the prosecution of any individual who has been identified as being involved in criminal acts of any nature. Upon completion, the time record shall be attached to and made part of the incident report. As may be required, copies of each record shall be maintained inhouse with originals of such offered to law enforcement authorities.

tort: A willful or negligent wrong committed against a person by another; a private or civil wrong or injury, other than a breach of contract, for which the court will provide a remedy in the form of an action for damages.

toxins/toxin agents: Poisonous by-products of living organisms used or dispersed to cause illness or death.

USC: United States Code (federal law).

vicarious liability: Liability attributed to the wrongdoer and to his or her trainer, supervisor, or employer; or one who has delegated authority to the wrongdoer; or one who has all or some authority over the wrongdoer's actions. In essence, everyone in a chain of authority may be subject to a civil suit.

viral agents: Viruses selected as biological weapons because of their ability to produce illness and death. See also *Ebola* and *Marburg*.

WHO: World Health Organization, an agency of the United Nations.

willful tort: See intentional tort.

WMD: Abbreviation for *weapons of mass destruction*, which include nuclear, biological, or chemical weapons that can be used by terrorists to inflict major damage and casualties.

Bibliography

Alexander, David. 2002. *Conspiracies and Cover-Ups*. New York: Berkley Books.

Beahm, George. 2004. *Straight Talk about Terrorism*. Dulles, VA: Brassey's.

Black, Henry Campbell. 1999. *Black's Law Dictionary*. 7th ed. St. Paul, MN: West Publishing.

Bolinger, Matt, M.D. 2004. *Recognizing and Treating Exposure to Anthrax, Smallpox, Nerve Gas, Radiation and Other Likely Agents of Terrorist Attack*. Boulder, CO: Paladin Press.

Clayton, Bruce D. 2002. *Life after Terrorism*. Boulder, CO: Paladin Press.

Couch, Dick. 2003. *The United States Armed Forces Nuclear, Biological and Chemical Survival Manual*. New York: Basic Books.

Denning, Dorothy. "Cyberterrorism." *Global Dialogue* Autumn (2002). A quarterly publication by Centre for World Dialogue, Nicosia, Cyprus. www.worlddialogue.org.

Disaster Preparedness. 2003. Washington, DC: Small Business Administration.

Elshtain, Jean Bethke. *Just War against Terror: The Burden of American Power in a Violent World*. New York: Basic Books.

Hill, Gerald N., and Kathleen Thompson Hill. 2002. *The People's Law Dictionary*. New York: MJF Books.

Lane, Fred H. 2004. *The Concerned Citizen's Guide to Surviving Nuclear, Biological, and Chemical Terrorist Attack*. Boulder, CO: Paladin Press.

Manley, Anthony D. 2004. *The Retail Loss Prevention Officer: The Law and the Fundamental Elements of Retail Security*. Upper Saddle River, NJ: Prentice-Hall.

New York State Mental Hygiene Law.

New York State Penal Law.

Nudell, Mayer, and Norman Antokol. *Handbook for Effective Emergency and Crisis Management*. Lexington, MA: Lexington Books, 1988. Reprinted by Mayer Nudell, 1999.

Parfomak, Paul W. "Guarding America: Security Guards and U.S. Critical Infrastructure Protection." A CRS report for Congress, November 12, 2004. Washington, DC: Congressional Research Service, The Library of Congress, Resources, Science and Industry Division.

Posner, Richard A. *Catastrophe:* 2004. *Risk and Response*. New York: Oxford University Press.

Project Response: The Oklahoma City Tragedy. A report provided in 1995 by the International Association of Chiefs of Police, Alexandria, VA.

"Workplace Violence Awareness and Prevention." An informational and instructional package for use by employers and employees, Long Island Coalition for Workplace Violence Awareness and Prevention, Mineola, NY, February 1996.

Index

C

F

G

Q

R

S

T

U

V

W

For Product Safety Concerns and Information please contact our EU representative GPSR@taylorandfrancis.com Taylor & Francis Verlag GmbH, Kaufingerstraße 24, 80331 München, Germany

Printed and bound by CPI Group (UK) Ltd, Croydon, CR0 4YY
08/05/2025
01864456-0001